HEAT AND MASS TRANSFER BY COMBINED FORCED AND NATURAL CONVECTION

The Institution of Mechanical Engineers

HEAT AND MASS TRANSFER BY COMBINED FORCED AND NATURAL CONVECTION

A Symposium arranged by the
Thermodynamics and Fluid Mechanics Group
of the Institution of Mechanical Engineers
15th September 1971

1 BIRDCAGE WALK · WESTMINSTER · LONDON · SW1H 9JJ

ISBN 0 85298 093 0
Library of Congress Catalog Card Number 79–184585

CONTENTS

Heat and Mass Transfer

A SYMPOSIUM was held in the Assembly Hall, University of Manchester, Owens Park, on the 15th September 1971. It was sponsored by the Thermodynamics and Fluid Mechanics Group of the Institution. The conference was opened by Dr F. N. Furber, M.Sc.Tech., Ph.D., C.Eng., M.I.Mech.E., and 68 delegates registered to attend. The planning panel consisted of Dr F. N. Furber (Chairman), Dr D. Chisholm and Mr A. J. Glasspoole.

C112/71

COMBINED NATURAL AND FORCED LAMINAR CONVECTION FOR UPFLOW THROUGH HEATED VERTICAL ANNULI

K. SHERWIN*   J. D. WALLIS†

Laminar flow systems have been studied in which buoyancy aids the forced flow, causing the fluid velocities to increase near the heat transfer surface and decrease elsewhere. Developing flow analysis has been performed which provides profiles of axial and radial velocities. Experimental results have been obtained with water in a vertical annulus of radius ratio 3 with a uniformly heated inner surface, which confirm that the Nusselt numbers are increased. Flow visualization by dye injection was used to determine the onset of flow instability, and it was found, as for previous work in which the buoyancy opposed the forced flow, that the instabilities were associated with large radial components of velocity. The experimental points agreed reasonably with the theoretical curve for the onset of flow instability which was asymptotic to a constant value for fully developed flow.

INTRODUCTION

IN RECENT YEARS the problem of combined natural and forced laminar convection has received considerable attention. Most of the studies have concerned heat transfer within vertical systems. Hallman (**1**)‡, Scheele and Hanratty (**2**), and Rosen and Hanratty (**3**) provided solutions and experimental results for fully developed flows in circular tubes. Lawrence (**4**) extended the theory to developing flows. Solutions have also been provided for the parallel flat plate geometry for cases of fully developed flow (**5**) and developing flow (**6**). However, in practice many heat exchanger applications require consideration of an annular configuration. A typical application is that of nuclear reactors in which cylindrical members are placed axially in vertical coolant channels.

Previous studies by the authors have concerned fully developed flows (**7**) and developing flows (**8**) for flow down vertical, internally heated annuli. The present paper extends these studies to the case of upward flows.

With flows up vertical annuli the natural convection aids the main flow causing the flow velocities to increase near the heated surface. The resulting modification of the velocity profiles causes the heat transfer performance to vary from that predicted by pure forced laminar convection theory (**9**). It is reasonable to expect the increase in velocities near the heated surface to cause improvement in the heat transfer performance, and this has been confirmed for the case of heated circular tubes (**2**).

The MS. of this paper was received at the Institution on 31st March 1971 and accepted for publication on 10th May 1971. 23

* *Dept of Mechanical Engineering, University of Liverpool.*
† *Mechanical Engineering Dept, Lanchester Polytechnic, Coventry.*
‡ *References are given in Appendix 112.1.*

The only previous work found on combined forced and natural convection in annuli is by Beck (**10**), which was an experimental investigation for laminar upflow of oil in annuli with an outer surface heated to a constant temperature. Average Nusselt numbers were correlated for both concentric and eccentric annuli in terms of the Prandtl number, Reynolds number, radius ratio, and viscosity ratio. It was noted that heat transfer was enhanced at low flow conditions by superimposed natural convection, and that the influence of natural convection increased with increasing annulus width. Furthermore, it was noted that the heat transfer performance decreased with increasing annulus eccentricity. Unfortunately the results are of limited value as the magnitudes of the natural convection forces were not correlated.

Notation

- Gr Grashof number based on temperature difference $[= \beta g \rho^2 r_e^3 \theta / \mu^2]$.
- Gr_q Grashof number based on heat flux $[= \beta g \rho^2 r_e^4 q r_i / k \mu^2 (r_o + r_i)]$.
- k Thermal conductivity.
- Nu Nusselt number $[= 2 q r_e / k(t_i - t_{av})]$.
- P Pressure drop parameter $[= (dp/dx - \rho_0 g) r_e^2 / u_m \mu]$.
- q Heat flux.
- R Dimensionless radius $[r/r_e]$.
- R_i Inner dimensionless radius $[r_i/r_e]$.
- R_o Outer dimensionless radius $[r_o/r_e]$.
- Re Reynolds number based on radius $[= u_m \rho r_e / \mu]$.
- r_e Equivalent radius $[r_o - r_i]$.
- r_i Inner radius.
- r_o Outer radius.

T Dimensionless temperature [$= (t-t_o)/\theta$].
t_i Temperature at inner surface.
t_o Temperature at outer surface.
U Dimensionless axial velocity [$= u/u_m$].
u Axial velocity.
u_m Mean axial velocity.
V Dimensionless radial velocity [$= (u/u_m)Re$].
v Radial velocity.
X Dimensionless axial distance [$= (x/r_e)/Re$].
x Axial distance from start of heat transfer section.
β Coefficient of thermal expansion.
θ Temperature difference [t_i-t_o].
μ Viscosity.
ρ Density.

FULLY DEVELOPED FLOW THEORY

An analysis by Sherwin (11) covers the case of buoyancy aiding the flow for laminar flow within annuli. The analysis is based on the assumptions that:

(a) the flow is fully developed and symmetrical about the central axis;
(b) all fluid properties except density are constant, and the variation of density is only significant in the buoyancy term;
(c) the fluid is heated with uniform heat flux at the inner surface.

Natural convection effects are correlated with respect to the forced flow by Gr_q/Re.

The effect of natural convection is to increase the flow velocities near the heated surface. If the flow rate through the annulus is held constant, the velocity must decrease at the outer surface until, at a sufficiently large value of Gr_q/Re, the flow has zero velocity gradient at the outer surface and any further increase in heat input causes the flow to reverse in that region. For an annulus of radius ratio $R_o/R_i = 3$, the fully developed flow theory predicts the onset of reversed flow at a value $Gr_q/Re = 209$.

Theory indicates that buoyancy aiding flow improves the heat transfer performance. This trend is indicated by the theoretical curve presented later in the paper in Fig. 112.5, which shows the heat transfer performance for fully developed flow within an annulus of $R_o/R_i = 3$.

DEVELOPING FLOW THEORY

From a study of downflow within an annulus (7) it is apparent that flow reversal promotes large radial velocity components across the annulus. To provide a more realistic theoretical approach to the flow behaviour observed in practice the authors performed a theoretical study of developing flow for the case of buoyancy opposing the force flow (8). The analysis was based on the assumptions that:

(a) the flow is symmetrical about the central axis;
(b) as for the fully developed flow theory, the variation of density is only significant in the buoyancy term, and there is uniform heat flux at the inner surface;
(c) the second derivatives of axial velocity are only significant in the radial direction.

This theory can be extended to cover the case of buoyancy aiding flow by changing the sign of the buoyancy term in the momentum equation. Using the above assumptions and introducing non-dimensional forms of the variables, the momentum, continuity, and energy equations can be expressed in the forms:

$$U\frac{\partial U}{\partial X}+V\frac{\partial U}{\partial R}=\frac{1}{R}\cdot\frac{\partial U}{\partial R}+\frac{\partial^2 U}{\partial R^2}-P-\frac{Gr}{Re}T \quad (112.1)$$

$$\frac{\partial U}{\partial X}+\frac{\partial V}{\partial R}+\frac{V}{R}=0 \quad (112.2)$$

$$Pr\,U\frac{\partial T}{\partial R}+Pr\,V\frac{\partial T}{\partial R}=\frac{1}{R}\cdot\frac{\partial T}{\partial R}+\frac{\partial^2 T}{\partial R^2} \quad (112.3)$$

For the solution of all the variables one further equation is necessary in order to solve for P, the pressure drop parameter. It is most convenient to introduce a further continuity requirement of constant mass flow through the annulus:

$$\frac{R_o^2-R_i^2}{2}=\int_{R_i}^{R_o} UR\,\mathrm{d}R \quad (112.4)$$

The non-dimensional form of the momentum equation [equation (112.1)] expressed the effect of buoyancy in terms of Gr/Re, where the magnitude of the density change causing the buoyancy force is related to the temperature difference (t_o-t_i) at any axial position. Since the boundary condition at the inner surface is that of uniform heat flux it is more convenient to employ the modified form of the parameter Gr_q/Re, as applied to the fully developed flow theory, where the buoyancy force is related to the heat flux q:

$$\frac{Gr_q}{Re}=\frac{qr_ir_e}{k(r_o+r_i)\theta}\cdot\frac{Gr}{Re} \quad (112.5)$$

Heat transfer from the inner surface can be correlated by the Nusselt number based on the difference between the heated wall temperature and the bulk mean temperature of the fluid:

$$Nu = 2qr_e/k(t_i-t_{av}) \quad (112.6)$$

where the temperature difference is defined as

$$(t_i-t_{av})=\frac{\int_{R_i}^{R_o} U\theta(1-T)R\,\mathrm{d}R}{\int_{R_i}^{R_o} UR\,\mathrm{d}R} \quad (112.7)$$

By redefining the heat flux as $q = k\,\partial[\theta(1-T)]/\partial(r_eR)$, for the boundary R_i, local heat transfer results can be obtained using equations (112.6) and (112.7).

Solutions were obtained iteratively using an Elliot 803 computer from the starting-point that the flow entering the heat transfer section had a hydrodynamically developed velocity profile.

Computations were performed at each axial step in the following sequence. First, the finite difference form of the energy equation [equation (112.3)] was solved using a matrix inversion technique in order to find the values of T.

A value of P was then assumed and the finite difference form of the momentum equation [equation (112.1)] was solved for values of U, also using the matrix inversion technique. Validity of the values of U was checked by the continuity integral equation [equation (112.4)] and an iteration employed to find the correct value of P. This procedure was based on the assumption that the pressure drop parameter was constant over the cross-section at any particular axial position and could only vary directly with the axial velocities. Having found values of U, and knowing the boundary values of V at any particular axial step, all other values of V were computed directly from the finite difference form of the continuity equation [equation (112.2)].

Solutions were obtained for an annulus of outer to inner radius ratio $R_o/R_i = 3$ with 20 mesh spacings in the radial direction and an axial step width of $X = 0{\cdot}05$. Computation time for each axial step was of the order of 10 min on the Elliot 803, and this dictated the use of the very coarse axial mesh width. However, satisfactory solutions were obtained for several Gr_q/Re values.

Developing axial and radial velocity profiles for the case of $Gr_q/Re = 400$ are presented in Figs 112.1 and 112.2 for several steps up to $X = 3{\cdot}0$. Fig. 112.1 shows the development of the axial velocities and, for comparison, the fully developed isothermal axial velocity profile is presented as the case for $X = 0$. An increase in axial velocity is predicted due to the buoyancy aiding the forced flow with an associated slowing down near the outer surface. At $X = 3{\cdot}0$ the velocity gradient at the outer surface appears to be zero, judged within the limitations of the finite difference mesh spacings employed.

The developing radial velocity profiles for $Gr_q/Re = 400$ are presented in Fig. 112.2, the negative values indicating flow towards the inner heated surface. The magnitude of the radial velocities decreases with increasing axial distance

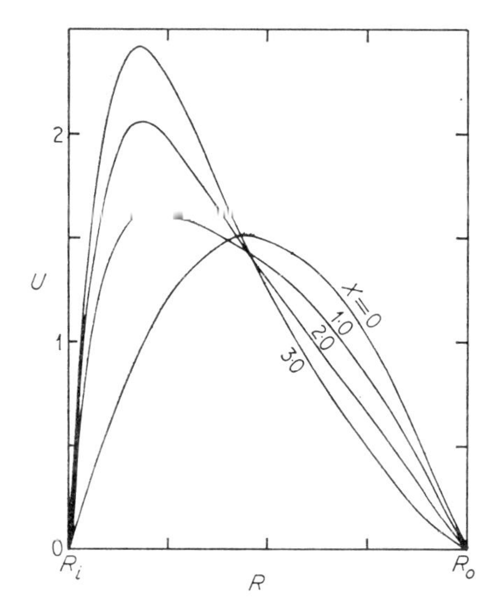

Fig. 112.1. Developing axial velocity profiles—$Gr_q/Re = 400$

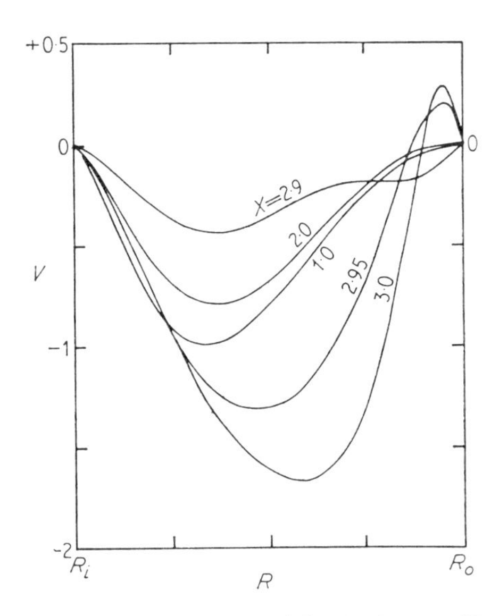

Fig. 112.2. Developing radial velocity profiles—$Gr_q/Re = 400$

X until $X = 2{\cdot}9$, above which there is an appreciable increase. In the previous study of buoyancy opposing flow (**8**) an increase in radial velocity was associated with reversed flows and caused flow to become unstable. In the present study the increase in radial velocity occurs near the point of zero axial velocity gradient ($X = 3{\cdot}0$) and can also be presumed to initiate unstable flow behaviour. Moreover, Fig. 112.2 shows a reversal in the radial flow near the outer surface above $X = 2{\cdot}9$, indicating a flow outwards in this region which, taken with the negative radial velocities nearer the axis, must indicate the onset of instabilities in the flow.

The onset of flow instabilities was predicted from the numerical analysis for several values of Gr_q/Re, and a plot Gr_q/Re against X is included in Fig. 112.3. The trend indicated by this curve agrees with that for similar theoretical results by Lawrence (**4**) for buoyancy aiding flow within a vertical circular tube. Furthermore, as X becomes large the value of Gr_q/Re required for the flow instabilities to initiate tends towards the value of 209 predicted by the fully developed flow theory for zero velocity gradient at the outer surface.

Results for the case of $Gr_q/Re = 400$ have not been presented for values of X larger than 3·0 since solutions were unobtainable from the numerical analysis above this value due to failure to converge on the required values of axial velocity. This computational limitation was noted near the point of zero velocity gradient at other values of Gr_q/Re, and tends to coincide with the predicted onset of unsteady flow behaviour.

EXPERIMENTAL FLOW BEHAVIOUR

Flow visualization experiments were performed using water in a vertical annulus of 19·1 mm inner and 57·2 mm outer diameter (i.e. $R_o/R_i = 3$), with uniform heating at

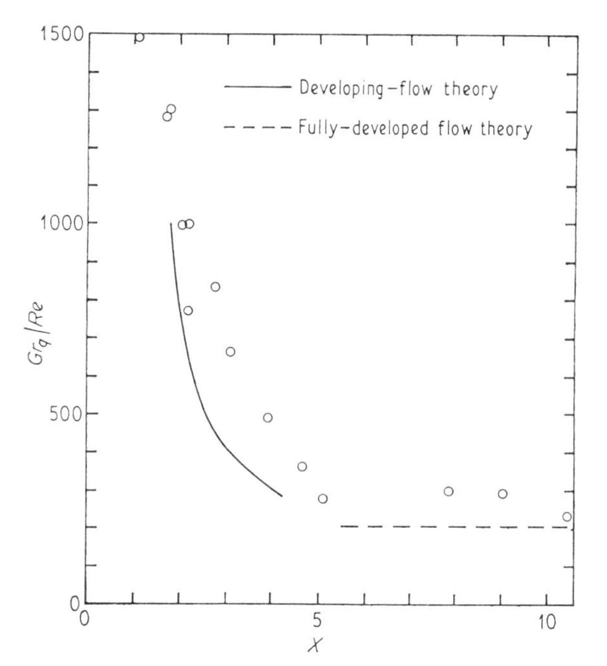

Fig. 112.3. Onset of flow instabilities

the inner surface. The heat transfer section had a length to equivalent radius ratio of 96 preceded by the calming section having a comparable ratio of 30. The rig was the same as that previously described (7) except for reversal of the flow direction.

Flow patterns were indicated by the injection of dye into the water stream. Several injection points were employed consisting of hypodermic needles set normal to the flow with the dye exit set midway between the inner and outer tubes of the annulus.

The flow could be directly observed within the heat transfer section through the transparent outer tube. Observation confirmed the developing flow theoretical predictions concerning the increase in radial flow near the point of zero velocity gradient at the outer surface, and Fig. 112.4 shows an example of the radial flow movements. It also shows that at those particular values of Gr_q/Re and X the flow had become unsteady, not only radially but also vertically, so that although the flow was upward there was some local downflow: the dye stream initially moved downwards before following the general radial flow pattern. No flow reversal was observed ahead of the unstable regime, providing good agreement with the predicted theoretical flow behaviour.

Experimental results for the onset of unstable flow are plotted on Fig. 112.3. These provide reasonable agreement with the theoretical curve, any differences between them probably being due to an asymmetry in the flow which was evident near the transition to the unstable region. Asymmetric flows could result from a difference in circumferential heating. The heater tube had a nominal wall thickness of 0·35 mm and measured difference in wall thickness of 0·025 mm. However, asymmetric flows were

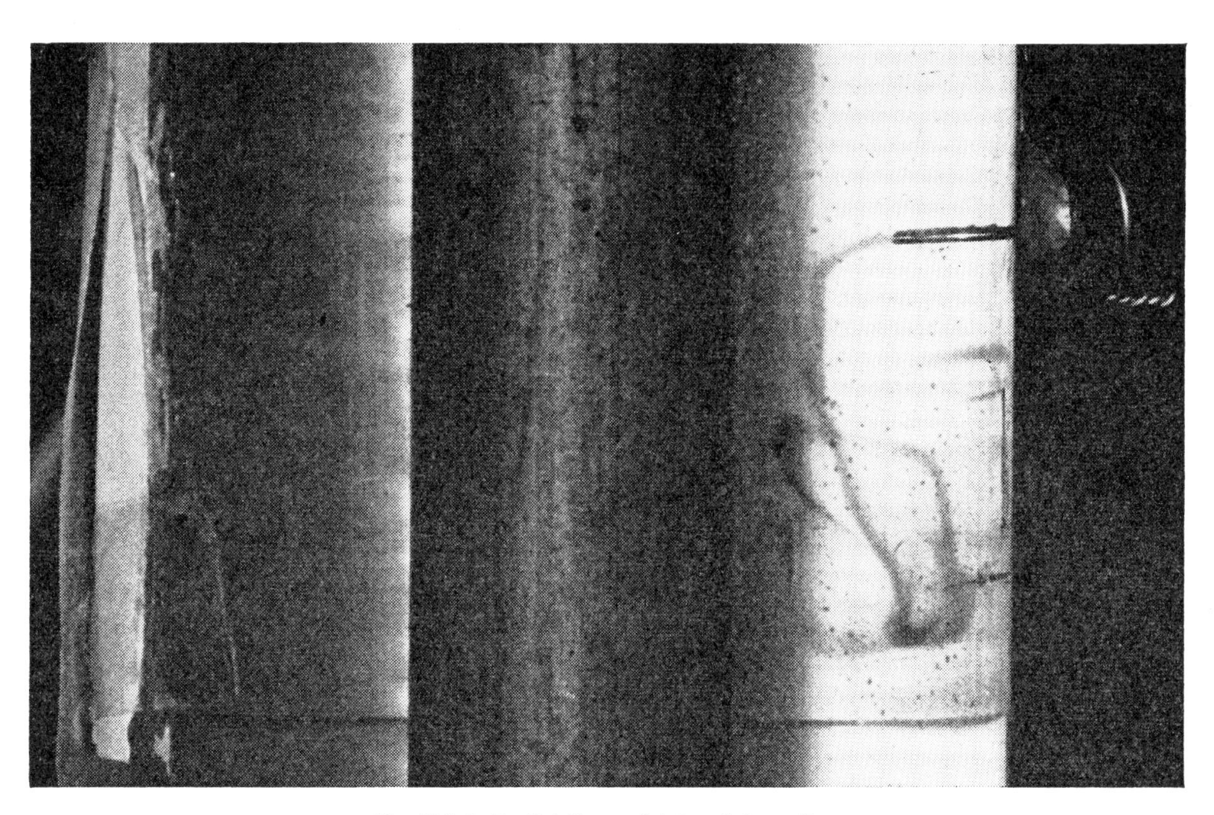

Fig. 112.4. Radial flow with local downflow

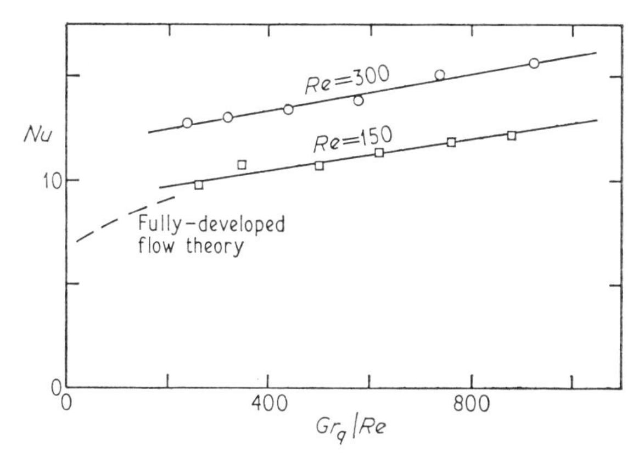

Fig. 112.5. Heat transfer results for upflow

not observed when using the same rig for downflow experiments (7).

EXPERIMENTAL HEAT TRANSFER

Experimental heat transfer results based on the heat flux as found from the power input to the test rig and on the temperature difference between the mean bulk fluid temperature and the mean heated surface temperature are shown in Fig. 112.5. Results are plotted for two flow rates equivalent to Reynolds values of 150 and 300 respectively. Owing to the comparatively short isothermal calming entrance length results were not obtained for higher *Re* values.

The results indicate the dependence of heat transfer performance on Reynolds number, confirming the previous experimental results of Scheele and Hanratty (**2**) for heat transfer in a vertical circular tube.

The results as presented in Fig. 112.5 apply to the stable flow regime. However, it would appear from the previous work by Scheele and Hanratty (**2**) that there is no marked change in the heat transfer performance during the transition to unstable flow behaviour.

At $Re = 150$ the value of X approaches that for fully developed flow, and the experimental heat transfer results are near those predicted by theory for fully developed flow.

CONCLUSIONS

With buoyancy aiding the main flow, the developing flow theory indicates that the radial flow motion becomes unstable as the velocity gradient approaches zero at the outer unheated surface. Unstable flow occurs when the parameter Gr_q/Re exceeds a certain critical value, which gradually reduces with developing flow until reaching a value appropriate to fully developed flow. Fig. 112.3 shows reasonable agreement between theory and experiment over a wide range of developing flow, with the curve asymptotic to a value for Gr_q/Re of 209 for fully developed flow in an annulus of radius ratio 3.

Experimental heat transfer results vary with Reynolds number, confirming the trend indicated by the developing flow theory. At the smallest value of $Re = 150$ the experimental results approach those predicted by theory for fully developed flow.

APPENDIX 112.1

REFERENCES

(**1**) Hallman, T. M. 'Combined forced and free convection in a vertical tube', Ph.D. thesis, Purdue University, 1958 (May).

(**2**) Scheele, G. F. and Hanratty, J. T. 'Effect of natural convection instabilities on rates of heat transfer at low Reynolds numbers', *A.I.Ch.E. Jl* 1963 **9** (No. 2), 183.

(**3**) Rosen, E. M. and Hanratty, J. T. 'Use of boundary-layer theory to predict the effect of heat transfer on the laminar-flow in a vertical tube with a constant temperature wall', *A.I.Ch.E. Jl* 1961 **7** (No. 1), 112.

(**4**) Lawrence, W. T. 'Entrance flow and transition from laminar to turbulent flow in vertical tubes with combined free and forced convection', M.I.T. Rept No. D.S.R. 9365, 1965.

(**5**) Rao, T. L. S. and Morris, W. D. 'Superimposed laminar forced and free convection between parallel plates when one plate is uniformly heated and the other is thermally insulated', *Thermodynamics and Fluid Mechanics Conv., Proc. Instn mech. Engrs* 1967–68 **182** (Pt 3H), 374.

(**6**) Savkar, S. D. 'Developing forced and free convective flows between two semi-infinite parallel plates', Paper NC.3.8, *4th Int. Heat Transfer Conf.*, Paris, 1970.

(**7**) Sherwin, K. and Wallis, J. D. 'A study of laminar convection for flow down vertical annuli', *Thermodynamics and Fluid Mechanics Conv., Proc. Instn mech. Engrs* 1967–68 **182** (Pt 3H), 330.

(**8**) Sherwin, K. and Wallis, J. D. 'A theoretical study of combined natural and forced laminar convection for developing flow down vertical annuli', Paper NC.3.9, *4th Int. Heat Transfer Conf.*, Paris, 1970.

(**9**) Lundberg, R. E., Reynolds, W. C. and Kays, W. M. 'Heat transfer with laminar flow in concentric annuli with constant and variable wall temperatures and heat flux', Stanford University Rept AHT-2, 1961.

(**10**) Beck, F. 'Wärmedbergung und Obuckverlust in senkreihten konzentrische und exzentrische Ringspalten bei erzwangener Strömung und frier Konvection', *Chemie-Ingr-Tech.* 1963 **35** (No. 12).

(**11**) Sherwin, K. 'Laminar convection in uniformly heated vertical concentric circular annuli', *Br. Chem. Engng* 1969 **13** (No. 11), 1580.

C113/71 INTERACTION BETWEEN A TURBULENT FREE CONVECTION LAYER AND A DOWNWARD FORCED FLOW

W. B. HALL* P. H. PRICE*

Experimental data are presented showing that heat transfer from a vertical plate by combined forced and free turbulent convection improves when the two components oppose each other. This contrasts with the case of laminar convection.

INTRODUCTION

THIS PAPER reports an extension of earlier work (1)† on mixed forced and free convection from a vertical plate. A turbulent free convection boundary layer was subjected to a vertical forced flow outside the layer; in the earlier work the forced flow was in the upward direction, whereas in the present case it is in the downward direction.

Whether a mixed convection situation is better regarded as one in which a forced flow is modified by buoyancy forces, or as a free convection exposed to an external forced flow, is a somewhat academic question. In this case the external forced flow was non-turbulent, and it is convenient to think of it as a modification to the velocity boundary condition at the outer edge of the free convection layer.

The conditions established in the experiment are met in a number of practical situations: a cavity with heated walls may be ventilated by a forced flow, and it then becomes important to establish whether this interacts significantly with the free convection. A closed cavity with heated and cooled walls may also generate large-scale flows that can be regarded as a perturbation on the free convection close to the walls. Theoretical studies of such interactions under non-turbulent conditions suggest that an upward external flow improves free convection at a heated surface, and that a downward flow reduces it. It was found, however, that the reverse is true in the case of an upward external flow and a turbulent free convection layer (1); under conditions in which the velocity of the external flow was similar in magnitude to the peak velocity in the free convection layer, heat transfer coefficients were reduced.

Earlier work by Eckert *et al.* (2) using a short vertical tube (of length/diameter ratio equal to 5) also gave decreased heat transfer when the forced flow was in the same direction as the free flow, and increased heat transfer when it opposed the free flow. In this case conditions were such that the free convection layer was turbulent.

The MS. of this paper was received at the Institution on 31st March 1971 and accepted for publication on 18th May 1971.
* *Simon Engineering Laboratories, University of Manchester.*
† *References are given in Appendix 113.1.*

Notation

Gr	Grashof number $[= g(\Delta T x^3/\bar{T}\nu^2)]$.
g	Gravitational acceleration.
k	Thermal conductivity of air.
Nu	Nusselt number $[= \alpha x/k]$.
Re	Reynolds number $[= u_s x/\nu]$.
T	Air temperature at a distance y from heated plate.
T_w	Temperature of heated wall.
T_0	Air inlet temperature.
T_1	Air temperature corresponding to $y = 0{\cdot}25$ m.
$\bar{T}$	$[=(T_w+T_0)/2]$.
u_m	Maximum velocity in free convection boundary layer.
u_s	Free stream velocity of forced flow.
x	Distance measured upwards from the bottom of the heated plate.
y	Distance from heated plate.
α	Heat transfer coefficient.
ν	Kinematic viscosity of air.

DESCRIPTION OF APPARATUS

The apparatus has been described in an earlier paper (1). Very briefly it consists of a vertical heated plate, 1·35 m high and 1·19 m wide, placed so as to form one wall of a vertical duct, as shown in Fig. 113.1. The top 0·75 m of the heated plate was split into horizontal strips, each fitted with a heater and thermocouples, and guarded by a heater at the rear to eliminate heat losses. The heat inputs to the individual strips could be adjusted to control the thermal boundary condition. In the present experiment the two lower branches of the duct were fitted with headers and connected to an extraction fan, thus enabling a downward forced flow to be produced in the duct. A perforated plate

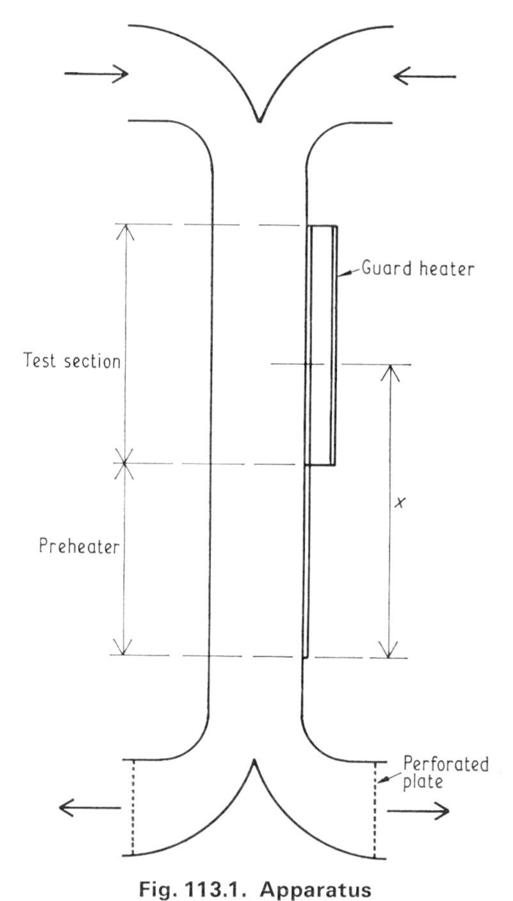

Fig. 113.1. Apparatus

of high flow resistance was inserted between the outlet branches and the headers to ensure a uniform rate of extraction across the width of the duct. The bulk downward flow in the duct was measured by means of an orifice plate situated between the apparatus and the extraction fan. At each value of plate temperature, measurements were made with the headers removed (i.e. with no forced flow) for comparison with the normal free convection situation.

In an attempt to test the hypothesis that improved heat transfer with downward flow is a result of a higher level of turbulence in the free convection layer, measurements of the mean temperature and the temperature fluctuations were made across the layer. A hot wire anemometer, operated in the constant current mode, was used for this purpose. Wires of 25 μm and 10 μm diameter were used and gave substantially the same value for the r.m.s. temperature fluctuation, thus suggesting that the 25-μm wire had an adequate frequency response.

EXPERIMENTAL PROCEDURE

One of the difficulties experienced with the apparatus was the length of time required to attain steady conditions. This was eventually achieved by fitting an automatic voltage regulator to the heater supply and a thermostat and heater to maintain uniform ambient temperature in the laboratory. It was generally possible to achieve steady conditions in the course of 6–8 h.

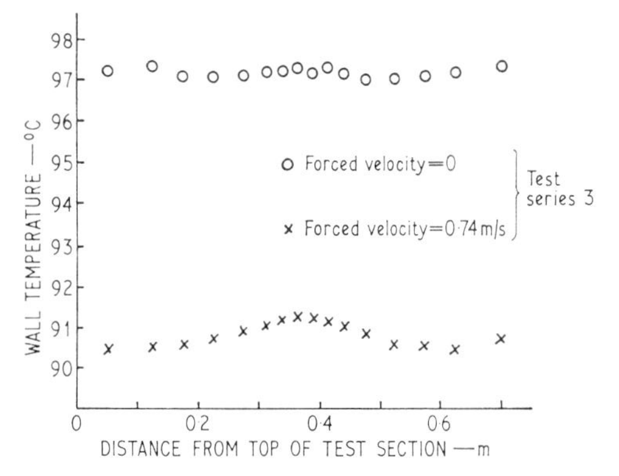

Fig. 113.2. Wall temperature distribution for free convection and for mixed convection

The results were found to depend on the vertical temperature gradients established in the laboratory. Under some conditions a variation in temperature of several degrees was established from floor to ceiling, and this resulted in a measurable change in heat transfer from the plate. By installing several 'chimneys' fitted with small fans drawing air from close to the ceiling of the laboratory and discharging it close to the floor, this temperature gradient was virtually eliminated.

Because of the time required to establish equilibrium, heat inputs to the sections of the plate were set to give a uniform wall temperature for the case of zero forced flow, and were then kept at the same value as the forced flow was introduced. Thus the conditions for forced flow are only approximately those of constant wall temperature (see Fig. 113.2).

Temperature fluctuations in the air were measured by feeding the output from the hot wire anemometer to a r.m.s. voltmeter with a low-frequency cut-off at about 0·1 Hz.

EXPERIMENTAL RESULTS

Heat transfer measurements

Fig. 113.2 shows a typical distribution of wall temperature over the height of the test section for zero forced flow and for the maximum forced flow. As mentioned above, the wall temperature was adjusted to be uniform for the case of zero forced flow; with a forced flow the variation did not exceed 4 per cent of the temperature difference between wall and air temperatures for the whole range of results. Heat flux distributions were found to be uniform to within ±3 per cent over the height of the test section when the temperature had been made uniform for the zero flow experiment in a test series; the heat inputs were then maintained at the same values for the forced flow experiments in that series. Because of the uniformity of wall temperature and heat flux the results have been expressed in terms of average quantities over the whole of the test section, and are presented in Table 113.1 in this form. (See also Figs 113.3 and 113.4.)

Table 113.1

Test series		Forced flow velocity, m/s					
		0	0·28	0·39	0·55	0·64	0·74
1	Inlet air temperature, °C	24·8	31·0	29·5	32·8	32·8	31·0
	Temperature difference, degC	36·3	33·2	31·3	30·0	29·4	28·1
	Wall heat flux, W/m²	201	202	202	203	203	201
2	Inlet air temperature, °C	26·5	31·5	30·5	32·5	32·0	30·0
	Temperature difference, degC	50·6	46·8	45·2	43·1	41·7	40·5
	Wall heat flux, W/m²	306	307	307	307	311	312
3	Inlet air temperature, °C	31·3	34·0	33·0	33·0	34·6	34·0
	Temperature difference, degC	66·0	62·3	61·0	59·3	57·7	56·8
	Wall heat flux, W/m²	457	459	457	457	456	460
4	Inlet air temperature, °C	30·0	—	31·0	32·0	34·0	33·5
	Temperature difference, degC	85·0	—	78·3	75·8	74·3	73·0
	Wall heat flux, W/m²	635	—	631	637	650	636

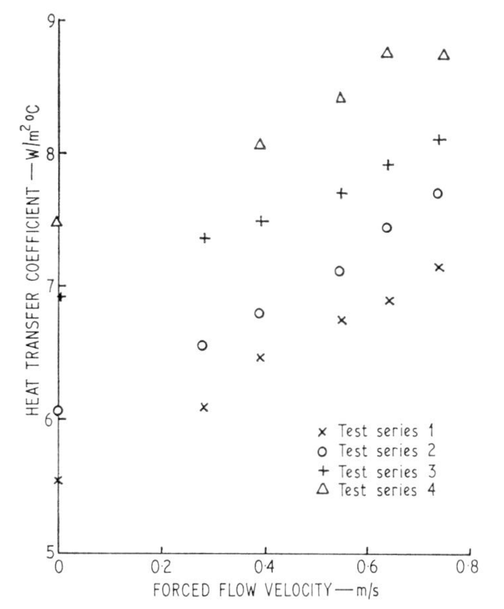

Fig. 113.3. Effect of forced flow on the heat transfer coefficient

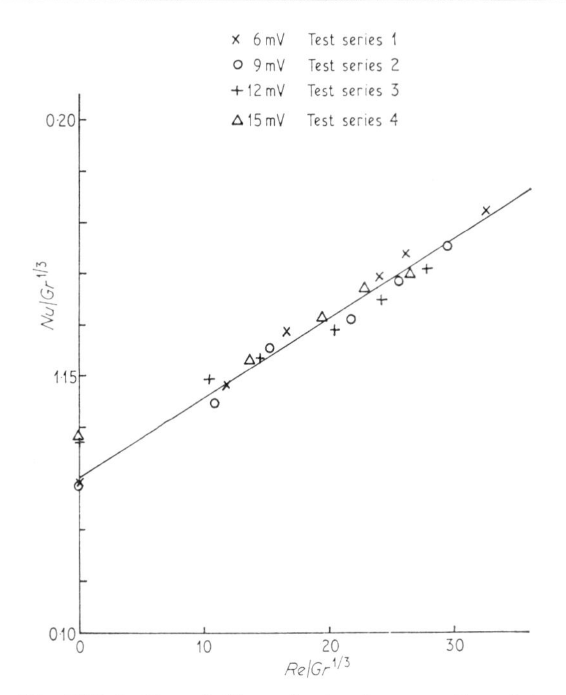

Fig. 113.4. Correlation of mixed convection data

Air temperature traverses

Traverses of the thermal boundary layer were made to determine distributions of time-mean temperature and temperature fluctuation across the boundary layer. Figs 113.5, 113.6, 113.7, and 113.8 show these distributions for the case of heat flux of 460 W/m²; Figs 113.5 and 113.6 on the one hand, and Figs 113.7 and 113.8 on the other, refer to vertical positions 0·15 m and 0·75 m respectively below the upper end of the heated plate. Figs 113.5 and 113.7 are distributions of time-mean temperature relative to the air inlet temperature, and Figs 113.6 and 113.8 are distributions of temperature fluctuations. On each figure are curves relating to zero forced flow and downward flows of 0·39 and 0·64 m/s. The temperature fluctuations (Figs 113.6 and 113.8) have been normalized by dividing the r.m.s. fluctuation by the temperature difference between the heated wall and the air at a distance of 0·25 m from the wall.

Accuracy of measurements

The estimated accuracy of the heat transfer results is ±5 per cent, the main limitation being the measurement of heat flux from the plate. The test section was guarded by heaters placed along the vertical edges and at the rear; however, non-uniform distribution of heaters on the rear guard plate gave rise to temperature variations of up to 10 degC over its surface. While the absolute accuracy is limited in this manner, the relative accuracy for tests with and without forced flow is probably within ±2 per cent; it was certainly possible to repeat measurements to better than ±1 per cent. As mentioned earlier, an important factor in achieving repeatability was the elimination of temperature gradients in the laboratory. This was particularly important in the experiments with no forced flow,

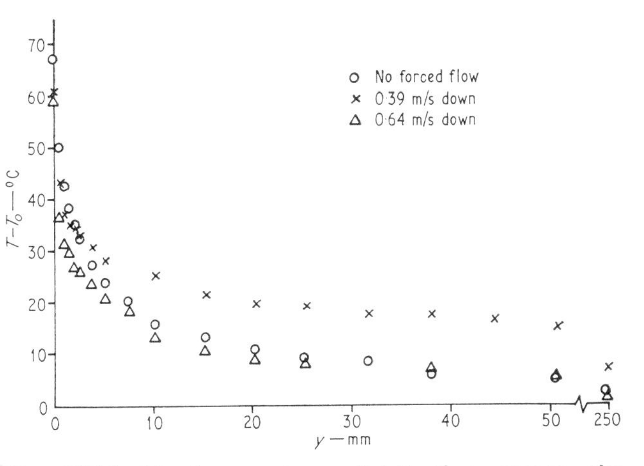

Fig. 113.5. Air temperature distribution across the thermal layer at a station 0·15 m below the top of the test section

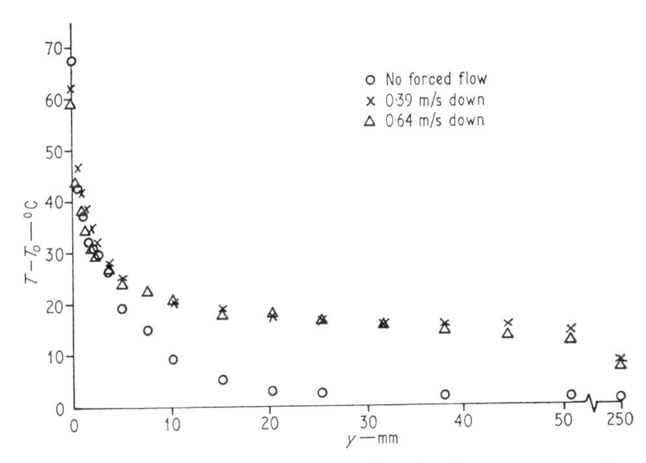

Fig. 113.7. Air temperature distribution across the thermal layer at a station 0·75 m below the top of the test section

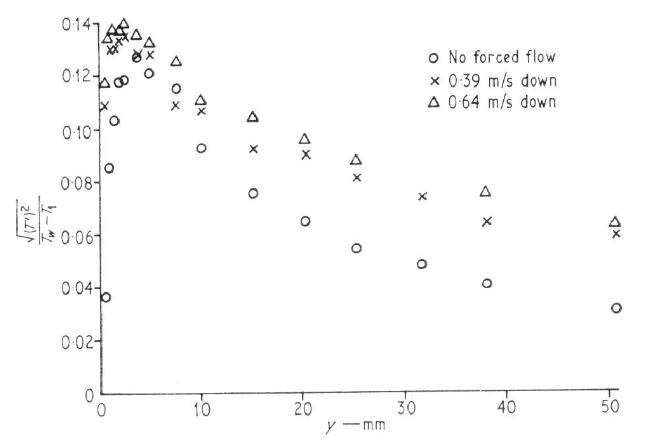

Fig. 113.6. Normalized air temperature fluctuations across the thermal layer at a station 0·15 m below the top of the test section

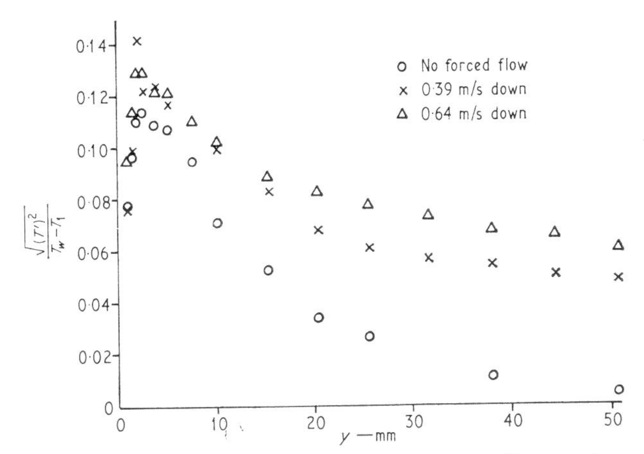

Fig. 113.8. Normalized air temperature fluctuations across the thermal layer at a station 0·75 m below the top of the test section

when a temperature 'inversion' of several degrees formed unless steps were taken to introduce vertical mixing.

The air temperature measurements across the boundary layer are probably accurate to ± 3 degC. The anemometer was used as a resistance thermometer and was calibrated against a mercury-in-glass thermometer over the temperature range 30–100°C. This calibration was performed with the wire immersed in oil, and was checked at one temperature with the wire in still air (to ensure that the current used was not sufficient to heat the wire significantly above the surrounding air). The accuracy of this calibration was estimated as ± 1 degC—the remainder of the error being incurred in estimating the mean of the fluctuating signal. This estimate was made by sampling and averaging the signal. The measurements of air temperature fluctuation have a frequency range that is limited on the one hand by the upper frequency limit of the wire ($\sim$ 20 Hz) and on the other by the lower frequency of the r.m.s. voltmeter used to measure the signal ($\sim$0·1 Hz). As mentioned above, the use of the smaller wire did not significantly affect the result, and it was concluded that the 25-μm wire had an adequate response. Both the r.m.s. and the mean of the signal were checked in a few cases by sampling at a frequency of 10 Hz over a period of 40 min; the results agreed with those obtained in the manner described above within ± 5 per cent in the case of the r.m.s. fluctuation and $\pm 1·5$ degC in the case of the mean temperature. It must be stressed that the temperature and temperature fluctuation traverses were intended primarily to give a qualitative indication of the effect of the forced flow on the heat transfer process. Regarded in this light, the accuracy achieved is adequate.

ANALYSIS AND DISCUSSION OF RESULTS

Heat transfer coefficient

The data of Table 113.1 are shown in Fig. 113.3 in the form, for each test series, of average heat transfer coefficient as a function of forced flow velocity. The heat transfer coefficient for the maximum forced flow velocity is about

28 per cent higher than that for zero flow in the case of test series 1, and about 19 per cent higher in the case of test series 4. These increases may be put into perspective by comparing the forced flow velocities with the peak velocities in the free convection layer (for zero flow). The latter velocities, estimated from the measurements of Cheesewright (3), can be expressed in the following form:

$$u_m \approx 0{\cdot}3\sqrt{(Gr)}\cdot\frac{\nu}{x} \quad . \quad . \quad (113.1)$$

The values corresponding to the four test series, based on a value of x equal to the distance from the bottom of the heated plate to the mid-point of the test section, are:

Test series	u_m, m/s
1	0·30
2	0·35
3	0·40
4	0·44

It appears, therefore, that a significant effect on heat transfer is obtained when the external forced flow velocity is of the same order as the peak velocity in the free convection layer. A similar result was found in the case of an upward external flow (1); in that case the heat transfer coefficient reached a minimum when the two velocities were approximately equal.

A comparison may be made with the heat transfer coefficient that would result from forced convection alone. The Reynolds number at the bottom of the test section (0·75 from the start of the boundary layer, since the flow is downwards and the test section is 0·75 m high) at a free stream velocity $u_s = 0{\cdot}74$ m/s and a temperature of 30°C would be $3{\cdot}5\times10^4$. The boundary layer would therefore be laminar, and the average heat transfer coefficient over the test section would be 3·7 W/m^2 degC. Referring to Fig. 113.3 and Table 113.1, we see that for test series 1 (with a temperature difference, $T_w - T_1$, ranging from 36 to 28 degC) the observed heat transfer coefficient varies from about 5·5 W/m^2 degC for zero forced flow to 7·2 W/m^2 degC for the maximum forced flow velocity of 0·74 m/s. The observed heat transfer coefficient for downward flow is thus seen to be greater than either the pure free convection or the pure forced convection heat transfer coefficients.

Dimensionless presentation of results

In the following section the physical properties of the Grashof and Nusselt numbers are evaluated at a temperature equal to the average of the wall temperature and the air inlet temperature. The kinematic viscosity in the Reynolds number is evaluated at the inlet temperature.

The uniformity of the heat transfer coefficient, for zero forced flow, over the height of the test section is consistent with other data for turbulent free convection: it implies a relationship of the form

$$Nu_0 \propto Gr^{1/3} \quad . \quad . \quad . \quad (113.2)$$

so that the characteristic dimension x cancels out, leaving the heat transfer coefficient independent of position.

Fig. 113.2 shows that even with a forced flow the heat transfer coefficient is very nearly independent of position on the test section. Fig. 113.3 suggests a linear variation of heat transfer coefficient with the forced flow velocity, u_s. This implies a relationship between the Nusselt number for zero forced flow, Nu_0, and that for mixed flow, Nu, of the form

$$Nu - Nu_0 \propto Re \quad . \quad . \quad . \quad (113.3)$$

thus preserving the independence of the heat transfer coefficient and the characteristic dimension, x. This form of presentation is adopted in Fig. 113.4, in which all the results are seen to agree reasonably well with the expression

$$Nu = 0{\cdot}13Gr^{1/3} + 0{\cdot}001\,56Re$$

or, if $Pr = 0{\cdot}7$,

$$Nu = 0{\cdot}147Ra^{1/3} + 0{\cdot}001\,56Re . \quad (113.4)$$

This may be compared, for $Re = 0$, with the expression for turbulent free convection from a vertical plate (4)

$$Nu = 0{\cdot}135Ra^{1/3} \quad . \quad . \quad (113.5)$$

It is important to place limitations on the applicability of equation (113.4) outside the range of the experimental data. No such limitation is needed as $Re \to 0$ since the equation is then in reasonably good agreement with that for pure free convection [equation (113.5)]. As Re increases, however, the equation does not approach the correct form of expression for forced convection, as it might be expected to do (i.e. Nu should become proportional to $Re^{0{\cdot}8}$ rather than Re). In other words, the lack of dependence of α on x in equation (113.4) should not be maintained as forced convection becomes dominant. A fairly conservative criterion might be to limit the forced velocity to, say, $3u_m$ where u_m is the peak velocity that would exist in the pure free convection layer. This criterion may be expressed, using equation (113.1), in the form:

$$Gr/Re^2 > 1 \quad . \quad . \quad . \quad . \quad (113.6)$$

(It should be noted that whereas the characteristic dimension cancels out in equations (113.2), (113.3), (113.4), and (113.5), it does not in the inequality (113.6); it is suggested that, as in equation (113.1), the characteristic dimension in (113.6) should be the height of the heated plate.)

A further limitation to the applicability of equation (113.4) is the fact that the Grashof number should be sufficiently high for the mixed convection layer to be turbulent. The value of Gr at the bottom of the test section in test series 1 was $8{\cdot}2\times10^8$. According to Cheesewright (3) this is barely high enough to ensure a turbulent free convection layer. However, the mixed convection layer was certainly turbulent even at the lowest forced flow velocity used in the experiments. It is interesting to note, therefore, that even when the conditions are such that both the pure free convection and the pure forced convection boundary layers would be expected to be laminar, the mixed convection layer for downward flow may be turbulent.

Finally, a comment must be made on the effect of the width of the duct in a direction normal to the heated surface. In planning the experiments it was intended that this dimension should be large enough to accommodate both the thermal boundary layer and a reasonable thickness of 'undisturbed' external flow. There is evidence that this was not completely achieved under certain conditions, and the matter will be discussed further in relation to the air temperature distributions.

Comparison with the work of Eckert

Eckert *et al.* (**2**), summarizing their results for downward flow in a short vertical pipe (length/diameter ratio of 5), state that 'in the mixed-flow regime the heat transfer coefficients are always larger than the larger of the calculated forced-flow or free-flow heat transfer coefficients. The measured value was found to be, at a maximum, approximately twice as large as the calculated value.' They give a limit equation (based on a 10 per cent increase in heat transfer coefficient) between the free flow and the mixed flow regimes, in the form

$$Re = 16{\cdot}1 Gr^{1/3} \quad . \quad . \quad . \quad (113.7)$$

A similar criterion may be obtained from equation (113.4). The ratio of the Nusselt number for mixed convection, Nu, to that for free convection, Nu_0, is:

$$\frac{Nu}{Nu_0} = \frac{0{\cdot}13 Gr^{1/3} + 0{\cdot}001\,56 Re}{0{\cdot}13 Gr^{1/3}} = 1 + 0{\cdot}012 \frac{Re}{Gr^{1/3}}$$

If we set $Nu/Nu_0 = 1{\cdot}10$ then the criterion corresponding to equation (113.7) becomes

$$Re = 8{\cdot}33 Gr^{1/3} \quad . \quad . \quad . \quad (113.8)$$

The forms of the criteria are thus similar; the difference in the numerical constant is presumably the result of the different shapes of the two systems. As is the case with the present data, Eckert did not find it possible to define the boundary between mixed convection and forced convection.

Air temperature distributions

It will be seen from Fig. 113.5 (which refers to a point close to the top of the test section) that for both zero forced flow and for a forced flow of 0·64 m/s the air temperature approaches the inlet temperature quite rapidly. At the intermediate flow, however, the air temperature remains significantly above the inlet temperature over the greater part of the duct. Flow visualization tests using smoke indicated that for a downward forced flow of 0·34 m/s the flow close to the wall was still in an upward direction (although there were occasional reversals), whereas for a flow of 0·64 m/s the flow was everywhere downwards. At the intermediate flow, therefore, there was some recirculation of heated air with the forced flow and this was probably responsible for the effect on the temperature distribution noted above.

The temperature distributions at the bottom of the test section (Fig. 113.7) follow a rather different pattern. As in Fig. 113.5, that for zero forced flow falls rapidly to the inlet temperature, but the distributions for both forced flows show a higher temperature at a distance from the heated wall. Even though the forced flow of 0·64 m/s is sufficient to suppress recirculation, the interaction between the forced and free flows appears to result in a rapid lateral growth in the thickness of the thermal layer, so that it completely fills the duct at the position of the lower traverse.

The measurements of temperature fluctuation (Figs 113.6 and 113.8) indicate that the general effect of a downward flow is to increase the level of the normalized r.m.s. fluctuation at all points across the thermal layer. This is consistent with the result of measurements made by Brown and Gauvin (**5**) using water in a vertical pipe. Temperature fluctuations have been used as an indication of the turbulence level simply because the measurement is relatively simple compared with the measurement of velocity. Simultaneous temperature and velocity fluctuations, very high levels of turbulence, and reversals of the mean flow make the latter measurement a very formidable task.

It is clear from the above comments that whereas the extreme conditions of a dominant free flow or a dominant forced flow give a fairly conventional 'boundary layer' type of flow, there are intermediate conditions where there is an important recirculation in the duct. It is tempting to think in terms of a stagnation point followed by a 'separation bubble' at these intermediate conditions, since the results certainly span the condition at which the flow at the wall is reversed. However, flow visualization gave no indication of a steady separation point; there were generally quite frequent reversals of velocity close to the wall over the whole of the test section when the forced flow was set at an appropriate level. It is surprising, in view of the foregoing remarks, that the heat transfer coefficient remains relatively uniform over the test section, and varies quite smoothly as the conditions are varied from a pure free convection situation to a mixed convection in which all upward flow has been suppressed.

CONCLUSIONS

The following conclusions may be drawn from the work:

(1) The interaction of a downward forced flow with a turbulent free convection layer on a heated vertical plate results in an increase in the heat transfer coefficient. Over a limited range of Grashof and Reynolds numbers, the heat transfer rate is given by equation (113.4).

(2) The work is in qualitative agreement with that of Eckert *et al.* (**2**) in defining the boundary between free and mixed convection in terms of the parameter *Re*.

(3) Measurements of temperature across the thermal layer indicate that at intermediate values of the forced flow hot air from the plate flows upwards and recirculates with the forced flow; at higher forced flows the upward flow is suppressed. Throughout the range of conditions, however, the heat transfer coefficient remains uniform over the surface and varies linearly with the velocity of the forced flow.

ACKNOWLEDGEMENT

The authors wish to acknowledge the assistance of Mr C. Brindley in the construction and operation of the apparatus.

APPENDIX 113.1

REFERENCES

(1) HALL, W. B. and PRICE, P. H. 'Mixed forced and free convection from a vertical heated plate to air', Paper NC.3.3, *Proc. 4th Int. Heat Transfer Conf.*, Paris, 1970.

(2) ECKERT, E. R. G., DIAGUILA, A. J. and CURRAN, A. N. 'Experiments on mixed free and forced convective heat transfer connected with turbulent flow through a short tube', *NACA Tech. Note 2974* 1953.

(3) CHEESEWRIGHT, R. 'Turbulent natural convection from a vertical plane surface', *J. Heat Transfer* 1968 (February).

(4) KUTATELADZE, S. S. *Fundamentals of heat transfer* 1963, 293 (Arnold).

(5) BROWN, C. K. and GAUVIN, W. H. 'Temperature profiles and fluctuations in combined free and forced convection flows', *Chem. Engng Sci.* 1966 **21**, 961.

C114/71 COMBINED FREE AND FORCED CONVECTION EFFECTS IN FULLY DEVELOPED LAMINAR FLOW IN HORIZONTAL TUBES

A. LICHTAROWICZ*

The effect of free convection on the uniform heat flux problem in tubes with fully developed laminar flow has often been examined theoretically. In this paper the problem is examined experimentally. The apparatus used and the techniques employed are briefly described. The results are presented as the variation of Nusselt number with the product of Reynolds and Rayleigh numbers. As the flow and the heat flux are reduced, the results tend to the constant value of Nusselt number predicted by simple theory. Temperature profiles were measured in the fully developed thermal flow region. The symmetry of the temperature profile in the vertical plane is distorted by the natural convection effects, whereas the profile in the horizontal plane is virtually unaffected.

INTRODUCTION

DURING THE LAST FEW YEARS a number of papers have been published on combined forced and free convection in fully developed laminar flow in horizontal tubes. Newell and Bergles (1)† provide an extensive list of references on the subject. All the work published was mainly concerned with the range where natural convection effects were large and the results deviated substantially from the now classical solution of the constant heat flux problem where $Nu = \frac{48}{11}$ and is independent of Reynolds number. As the heat input rate is increased from zero, the fluid near the wall of the tube heats up more than the fluid further inside. As a consequence, the density decreases and natural convection effects begin to act. Two upward currents are set up along the side walls. These currents meet at the top where they join to travel downwards along the vertical diameter. Thus natural convection forms two vortices whose axes lie along the tube and are symmetrically displaced on both sides of the pipe centre-line. The motion formed by the superposition of the forced flow along the tube and these two natural convection vortices result in the formation of two spiralling flows. The spiralling is symmetrical about the vertical plane. As the result of the natural convection currents, the heat transfer from the walls to the fluid is increased and the Nusselt number is no longer constant.

Some time ago two projects (2) (3) were run with the object of demonstrating these effects to students. The results obtained are of some interest since they relate to the region where the natural convection currents are weak and only just begin to manifest themselves.

Notation

C_p Specific heat.
d Pipe diameter.
Gr Grashof number [$= (gd^3\beta\rho^2/\mu^2)(t_w - t_b)$].
g Gravitational acceleration.
h Heat transfer coefficient.
k Thermal conductivity.
l Axial length.
Nu Nusselt number [$= hd/k$].
Pr Prandtl number [$= C_p\mu/k$].
Ra Rayleigh number $\left[= \frac{g\beta d^4\rho}{16\alpha\mu} \cdot \frac{dt_w}{dl}\right]$.
Re Reynolds Number [$ud\rho/\mu$].
t Temperature.
u Mean velocity.
α Thermal diffusivity [$= k/\rho C_p$].
β Bulk coefficient of thermal expansion.
μ Viscosity.
ρ Density.

Subscripts

b Bulk.
w Wall.

APPARATUS

The apparatus used was fully described in references (2) and (3) and is shown diagrammatically in Fig. 114.1.

The MS. of this paper was received at the Institution on 7th April 1971 and accepted for publication on 10th May 1971.
* *Dept. of Mechanical Engineering, University of Nottingham.*
† *References are given in Appendix 114.1.*

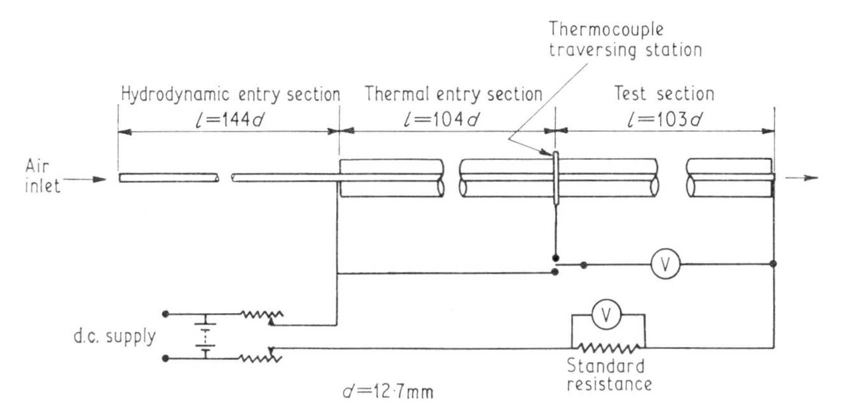

Fig. 114.1. Experimental arrangement

Briefly, the air flow provided by an electric blower is passed through a system consisting of air boxes, gas meter, rotameter, and a control valve. Air boxes are used to smooth out the flow pulsations caused by the gas meter and the rotameter indicates the stability of the flow. The air then passes through a long hydrodynamic entry tube ($l/d = 144$) and proceeds through a thermally insulating joint to the heated test section.

The test section, shown diagrammatically in Fig. 114.2, consists of two lengths of 18/8 stainless steel tube (total $l/d = 207$). The tube is 12·7 mm internal diameter and has a wall thickness of 1·59 mm. The tube is heated electrically by passing a heavy direct current through it. The basic electrical circuit is also shown in Fig. 114.1. Electrical power input was evaluated by measuring the voltage drop across the heater tube and across a standard resistance. Appropriate voltages were measured by the same potentiometer as was used for thermocouple measurements. Voltmeters were only used as indicators. The power input was regulated by the controls on the rectifier and by two rheostats.

The heater tube is placed inside a large-diameter 760-mm copper tube in which it is supported at both ends (insulated supports) and at the centre. The outside tube is heated by four independently controlled strip foil heaters. The heating to the outside tube can be adjusted so that the temperatures of both the heater and the outside tubes are the same, and there is a linear temperature gradient along the main tube. The thermal losses from the main tube are thus reduced virtually to zero. The central support carries a thermocouple traversing gear. An iron–constantan thermocouple made from 0·193-mm diameter wire was threaded through two of the supporting spokes (Fig. 114.2) and stretched from a traversing frame which could be moved by a micrometer head. An accurate temperature profile inside the tube could thus be obtained.

The whole test section could be rotated through 90° so that temperature traverses across a horizontal diameter could be taken.

The portion of the heated section upstream of the traversing station was used as a thermal inlet section ($l/d = 104$) while all the relevant measurements were made

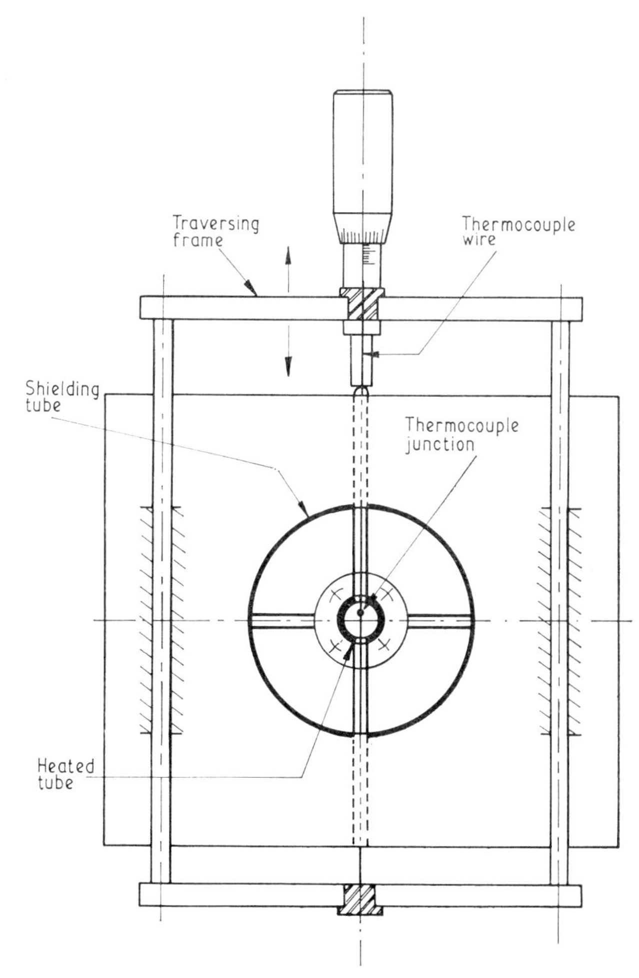

Fig. 114.2. Thermocouple traversing gear

on the downstream portion. A total of 48 iron–constantan thermocouples were used to measure temperatures. The thermocouples were attached, by Araldite, to the main tube in such a way that they were electrically insulated from it. The error introduced by this technique was estimated to be not more than 15 per cent of the temperature difference between the heated and the outer tubes,

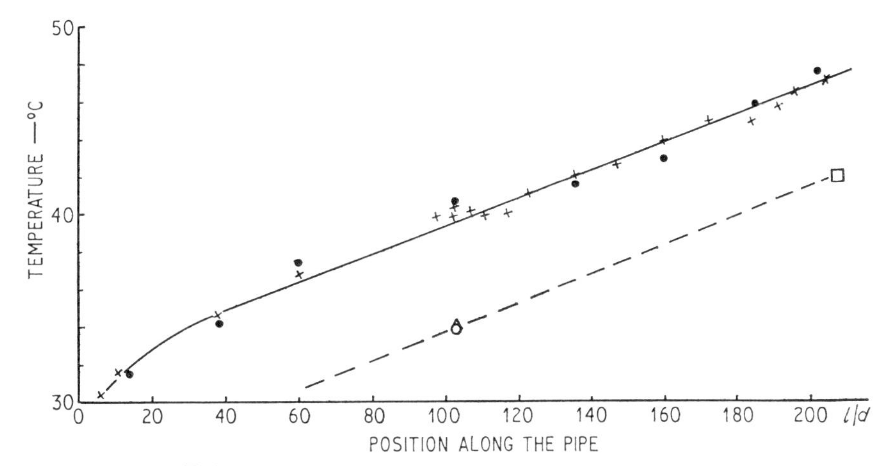

× Tube temperature.
● Outer shield temperature.
□ Measured air temperature.
△ Computed air temperature from power input.
○ Computed air temperature from measured temperature profile.

Test run 9, $Re \times Ra = 1362$.

Fig. 114.3. Typical temperature distribution along the rig

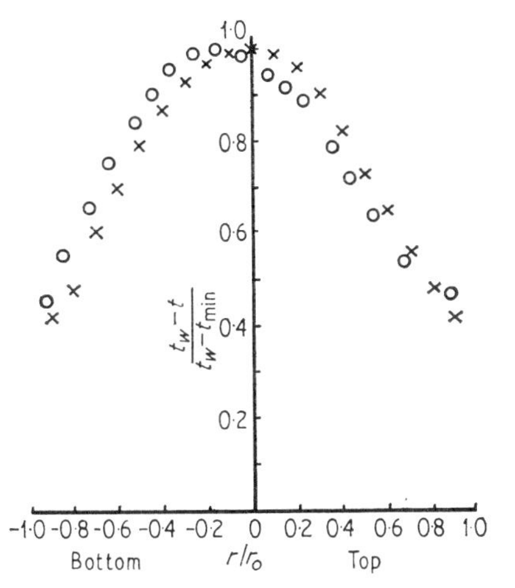

○ Vertical temperature distribution; test 9—$Re \times Ra = 1360$.
× Horizontal temperature distribution; test 8—$Re \times Ra = 1320$.

Fig. 114.4. Typical temperature profiles

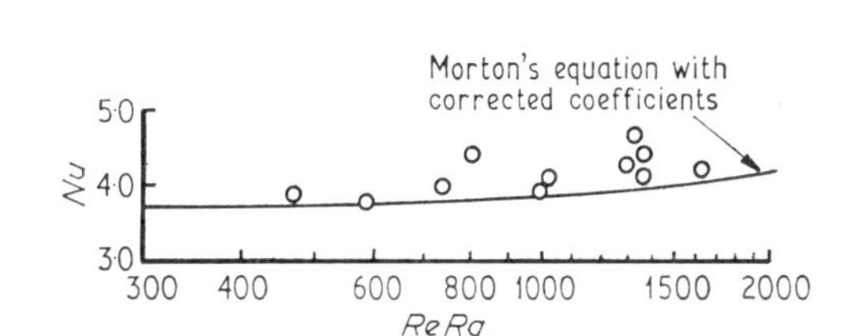

Fig. 114.5. Variation of Nusselt number with product of Reynolds and Rayleigh numbers

Table 114.1

Test run	Input power, W	Output power, W	Overall air temperature rise, degC	$t_w - t_b$, degC	Re	Ra	Gr	Nu	$Re \times Ra$
1	2·22	2·08	20·6	1·9	530	0·885	415	4·42	470
2	2·75	2·64	20·3	2·4	680	0·867	520	4·38	590
3	3·34	3·32	21·4	3·0	820	0·90	667	4·48	740
4*	3·72	3·53	21·8	3·25	840	0·96	737	4·70	803
5*	4·30	4·28	22·6	4·0	990	1·00	905	4·47	990
6	4·75	4·22	20·1	4·0	1200	0·84	840	4·54	1010
7	5·65	5·63	21·6	4·9	1390	0·93	1090	4·63	1300
8*	5·55	5·33	18·45	4·6	1520	0·87	1100	4·83	1320
9	6·4	6·11	18·7	5·8	1630	0·84	1230	4·55	1360
10	7·57	7·40	19·5	6·6	2010	0·82	1450	4·60	1640
11	6·48	6·34	18·4	5·5	1820	0·75	1170	4·71	1360

* Indicates readings with apparatus turned through 90°.

and this difference was always less than 1 degC. The overall error introduced was therefore less than 0·15 degC. All the thermocouples and associated circuits were calibrated against a mercury-in-glass thermometer. All such thermometers used were N.P.L. calibrated. The outlet air temperature was measured by a mercury-in-glass thermometer mounted inside the main pipe. A flow spoiler was used to mix the flow so that the bulk temperature could be measured. Inlet air temperature was measured at the start of the heated section by thermocouples. This temperature was the same as the air box temperature upstream of the hydrodynamic inlet section.

The bulk temperature of the air at the centre of the heated section could be calculated from the heat input to the test section and the overall air temperature rise. It could also be approximately calculated from the temperature profile at the centre section. In the absence of the measured velocity profile, an approximation was made in assuming the parabolic velocity profile.

The accuracy of the air flow measurement was within ± 1 per cent.

EXPERIMENTAL RESULTS

Considerable difficulty was experienced with the adjustment of outside tube strip heaters to give a linear temperature gradient along the heater tube, and when this difficulty was surmounted the testing was easy but slow. Typical results are shown in Figs 114.3 and 114.4 and the summary of all results is given in Table 114.1 and Fig. 114.5. The effectiveness of the thermal insulation of the heater tube is demonstrated by the close agreement between the measured electric power input and the heat carried away by the air.

From Fig. 114.5 it can be seen that the agreement between measured and calculated Nusselt numbers is good. The calculation was based on Morton's work with numerical constants (4), corrected by Mori and Futagami (5).

Natural convection currents begin to have a noticeable effect at $Re \times Ra$ values as low as 740. This is also noticeable from the vertical plane temperature profiles where the maximum point is displaced downwards. A typical temperature profile is shown in Fig. 114.4. The downward displacement is approximately proportional to the increase in Nusselt number. For comparison, a temperature across a horizontal diameter is also shown in the same figure. This profile is symmetrical about the centre-line. It was also noticed that the temperatures at the bottom of the heated tube were slightly lower than at the top, indicating that the natural convection currents affected the circumferential temperature distribution. The differences recorded were of the order of 0·5 degC.

CONCLUSIONS

It has been shown that Morton's approximate solution is valid up to the maximum $Re \times Ra = 1700$ tested, and that the natural convection effects can be detected at $Re \times Ra$ values as low as 740.

ACKNOWLEDGEMENTS

The author would like to express his gratitude to S. Dakarananda and P. J. Finch, honours degree candidates at the University of Nottingham, for the experimental work described in this paper. The facilities granted by Professor A. G. Smith, Head of the Department of Mechanical Engineering, are gratefully acknowledged.

APPENDIX 114.1

REFERENCES

(1) NEWELL, P. H. and BERGLES, A. E. 'Analysis of combined free and forced convection for fully developed laminar flow in horizontal tubes', Am. Soc. Mech. Engrs Paper No. 69-HT-39.

(2) FINCH, P. J. 'Heat transfer to laminar flow in a tube', B.Sc. thesis, Dept of Mechanical Engineering, University of Nottingham, 1963.

(3) DAKARANANDA, S. 'Heat transfer to laminar flow in a horizontal tube', B.Sc. thesis, Dept of Mechanical Engineering, University of Nottingham, 1965.

(4) MORTON, B. R. 'Laminar convection in uniformly heated pipes at low Reynolds numbers', *Q. Jl Mech. appl. Math.* 1959 **12**, 410.

(5) MORI, Y. and FUTAGAMI, K. 'Forced convective heat transfer in uniformly heated horizontal tubes', *Int. J. Heat Mass Transfer* 1967 **10**, 1801.

C115/71

COMBINED CONVECTION IN VERTICAL TUBES

M. W. COLLINS*

Laminar combined convection in vertical tubes is treated on the basis of a developing flow model, and the basic differential equations are rewritten in finite difference form and solved by a computer program in Fortran. In addition to the usual heat transfer calculations, this enables the developing velocity and temperature fields to be studied. The four main alternatives (of upward or downward flow with heating or cooling) are investigated, together with the wall boundary condition of constant heat flux or temperature. The behaviour of air and water forms a useful comparison in view of their different density and viscosity dependence on temperature.

INTRODUCTION

COMBINED (OR MIXED) CONVECTION arises when the normal modes of convection, viz. natural and forced, are simultaneously present. A practical instance of this is in the emergency cooling of nuclear reactors, where fault conditions result in only a low forced flow, superimposed on the natural convection in the reactor core. Interest in combined convection has grown increasingly in recent years and a comprehensive literature survey is given in Collins (1)†. This problem is complex because of the large number of sometimes interacting parameters, including the relative direction of the natural and forced convections, the geometry of the arrangement, the flow condition, and the boundary conditions.

Analytical work dates from that of Martinelli and Boelter (2) who, in 1942, treated a very simplified one-dimensional model. With the passage of time, and in particular the advent of digital computers, the number of assumptions made gradually decreased. The most important was that of a fully developed (one-dimensional) model. This has never been really justified (although still used for certain situations, e.g. Iqbal and Stachiewicz (3)) and in fact a developing model has been attempted by Rosen and Hanratty (4) and Lawrence and Chato (5). The former give an approximate solution, assuming a velocity profile as a three-term power series in r^2, and a temperature profile as a five-term exponential series. The unknown coefficients were solved by substitution into the available equations or boundary conditions, leading to a set of non-linear differential equations which were integrated using a digital computer. Lawrence and Chato used a finite difference approximation with matrix inversion, and reliably predicted developing velocity and temperature fields for water, again using a digital computer. Experimental work, mainly for circular pipes, include only a few which shed light on this developing problem—in particular, Scheele and Hanratty (6) and reference (5) above. Both these are for water, the latter for upward heated flow with constant heat flux only. Scheele and Hanratty prefer the general validity of a fully developed flow field, while Lawrence and Chato concentrate wholly on a developing situation. The problem is difficult to discuss owing to the absence of experimental velocity and temperature data for the former.

The final aspect which should be mentioned is instability and transition to a form of turbulent flow. Again, references (5) and (6) contain useful data. Transition is defined in both as the condition for which fluctuations in fluid temperature were observed, the latter specifically as 0·05 degF. Analytical prediction of transition is also attempted. For upward heated flow, reference (4) achieves partial success with a zero centre-line velocity prediction, although this precedes transition. Lawrence and Chato make an empirical correlation based on velocity profile shape.

A comparative analytical investigation of the main parameters does not exist, although this would aid a basic understanding of the combined convection process. A laminar flow treatment, while having rather limited practical importance, does serve to achieve this understanding without the added complexity and, at present, partial uncertainty of a turbulent approach. In addition, it could well serve as a base for the latter.

Notation

A	Wall area.
C_p	Specific heat of fluid at constant pressure.
G	$[= R^3 g/\nu_A^2]$, dimensionless.

The MS. of this paper was received at the Institution on 31st March 1971 and accepted for publication on 17th May 1971. 34

* *Mechanical Engineering Dept, The City University, London.*

† *References are given in Appendix 115.2.*

g	Acceleration due to gravity.
k	Thermal conductivity of fluid.
M	Number of radial divisions.
m, n	Radial and axial positions.
P	Dimensionless pressure $[= (p-p_0)/\rho_m U_0^2]$.
Pr	Prandtl number $[= \mu_A(C_p/k)_m]$.
p	Pressure.
Q	Wall heat flow rate.
R	Tube radius.
Re	Reynolds number $[= v_z R\rho_m/\mu_A]$.
r'	Radial co-ordinate.
r	Dimensionless radial co-ordinate $[= r'/R]$.
T'	Temperature.
T	Dimensionless temperature $[= G.C_5(T'-T'_0)]$.
U	Dimensionless axial velocity $[= v_z/U_0]$.
V	Dimensionless radial velocity $[= v_r/U_0]$.
v_r, v_z	Radial and axial velocities.
z'	Axial co-ordinate.
z	Dimensionless axial co-ordinate $[= z'/R]$.
μ, ν	Dynamic, kinematic viscosities.
ρ	Density.

Subscripts

A	Property reference point, equations (115.5) and (115.6).
m	Mean value.
R	Wall condition.
r, z	Radial and axial directions.
0	Entrance condition.

BASIC EQUATIONS

As discussed in the Introduction, the model is two-dimensional, axial, and radial, circumferential effects being absent. It deals with steady, developing vertical, laminar flow in a circular tube.

Equations (115.1)–(115.4) are for the following principles, respectively: conservation of mass, momentum in axial and radial directions, and energy. The only assumptions made are that C_p and k are constant, and that viscous dissipation and compressibility effects are negligible. The variation of ρ is expressed in the buoyancy term of equation (115.2). In addition, bulk density changes (i.e. densities compared with the value at entry) are allowed for by using varying ρ in the integral continuity equation:

$$\frac{\partial v_r}{\partial r'}+\frac{v_r}{r'}+\frac{\partial v_z}{\partial z'} = 0 \quad . \quad . \quad . \quad (115.1)$$

$$\rho\left(v_r\frac{\partial v_z}{\partial r'}+v_z\frac{\partial v_z}{\partial z'}\right) = -\frac{\partial p}{\partial z'}\pm\rho g +\mu\left(\frac{\partial^2 v_z}{\partial z'^2}+\frac{1}{r'}\frac{\partial v_z}{\partial r'}+\frac{\partial^2 v_z}{\partial r'^2}\right) +2\frac{\partial \mu}{\partial z'}\cdot\frac{\partial v_z}{\partial z'}+\frac{\partial \mu}{\partial r'}\left(\frac{\partial v_r}{\partial z'}+\frac{\partial v_z}{\partial r'}\right) \quad . \quad (115.2)$$

(where $\pm\rho g$ holds for downward or upward flow respectively)

$$\rho\left(v_r\frac{\partial v_r}{\partial r'}+v_z\frac{\partial v_r}{\partial z'}\right) = -\frac{\partial p}{\partial r'}+\mu\left(\frac{\partial^2 v_r}{\partial r'^2}+\frac{1}{r'}\cdot\frac{\partial v_r}{\partial r'}-\frac{v_r}{r'^2}+\frac{\partial^2 v_r}{\partial z'^2}\right) +2\frac{\partial \mu}{\partial r'}\cdot\frac{\partial v_r}{\partial r'}+\frac{\partial \mu}{\partial z'}\left(\frac{\partial v_r}{\partial z'}+\frac{\partial v_z}{\partial r'}\right) \quad (115.3)$$

$$v_r\frac{\partial T'}{\partial r'}+v_z\frac{\partial T'}{\partial z'} = \frac{k}{\rho G_p}\left(\frac{\partial^2 T'}{\partial z'^2}+\frac{1}{r'}\cdot\frac{\partial T'}{\partial r'}+\frac{\partial^2 T'}{\partial r'^2}\right) \quad (115.4)$$

The viscosity and density variations with temperature are taken to be:

$$\mu = \mu_A(C_1+C_2T'-C_3T'^2) \quad . \quad (115.5)$$

and

$$\rho = \rho_A(C_4-C_5T'+C_6T'^2) \quad . \quad (115.6)$$

where $C_1 \ldots C_6$ are constants. In addition, the integral continuity equation may also be used for a numerical solution, viz.:

$$\pi R^2\rho_0 U_0 = \int_0^R 2\pi r'\rho v_z\, dr' \quad . \quad (115.7)$$

Boundary conditions

Entry condition:

$$v_z = U_0, \quad v_r = 0, \quad p = p_0, \quad T' = T'_0$$

Tube axis:

$$v_r = 0, \quad \frac{\partial v_z}{\partial r'} = 0, \quad \frac{\partial T'}{\partial r'} = 0$$

Wall flow condition:

$$v_z = 0, \quad v_r = 0$$

Wall thermal condition, either

(*a*) Constant heat flux:

$$\frac{Q}{A} = k\left(\frac{\partial T'}{\partial r'}\right)_R = \text{const}$$

or (*b*) constant temperature:

$$T' = T'_R = \text{const}$$

Dimensionless equations

The above equations may be made dimensionless using the parameters defined in the Notation. Equations (115.1)–(115.7) become (115.8)–(115.14) respectively:

$$\frac{\partial V}{\partial r}+\frac{V}{r}+\frac{\partial U}{\partial z} = 0 \quad . \quad . \quad (115.8)$$

$$V\frac{\partial U}{\partial r}+U\frac{\partial U}{\partial z} = -\frac{\partial P}{\partial z}\pm\frac{1}{Re^2}[G.C_{04}+T(C_{05}+C_{06}T)]\frac{\rho}{\rho_A} +\frac{1}{Re}\left[\{C_{01}+T(C_{02}+C_{03}T)\}\times\left(\frac{\partial^2 U}{\partial z^2}+\frac{1}{r}\cdot\frac{\partial U}{\partial r}+\frac{\partial^2 U}{\partial r^2}\right)+(C_{02}+2C_{03}T)\times\left\{2\frac{\partial T}{\partial z}\cdot\frac{\partial U}{\partial z}+\frac{\partial T}{\partial r}\left(\frac{\partial V}{\partial z}+\frac{\partial U}{\partial r}\right)\right\}\right] \quad (115.9)$$

$$V\frac{\partial V}{\partial r}+U\frac{\partial V}{\partial z}=-\frac{\partial P}{\partial r}+\frac{1}{Re}\left[\{C_{01}+T(C_{02}+C_{03}T)\}\right.$$
$$\times\left(\frac{\partial^2 V}{\partial r^2}+\frac{1}{r}\cdot\frac{\partial V}{\partial r}-\frac{V}{r^2}+\frac{\partial^2 V}{\partial z^2}\right)+(C_{02}+2C_{03}T)$$
$$\left.\times\left\{2\frac{\partial T}{\partial r}\cdot\frac{\partial V}{\partial r}+\frac{\partial T}{\partial z}\left(\frac{\partial V}{\partial z}+\frac{\partial U}{\partial r}\right)\right\}\right] \quad (115.10)$$

$$V\frac{\partial T}{\partial r}+U\frac{\partial T}{\partial z}=\frac{1}{Pr.Re}\left(\frac{\partial^2 T}{\partial z^2}+\frac{1}{r}\cdot\frac{\partial T}{\partial r}+\frac{\partial^2 T}{\partial r^2}\right) \quad (115.11)$$

$$\mu=\mu_A(C_{01}+C_{02}T+C_{03}T^2)\,. \quad (115.12)$$

$$\rho=\rho_A\left(C_{04}+\frac{C_{05}T}{G}+\frac{C_{06}T^2}{G}\right). \quad (115.13)$$

$$\frac{C_{04}}{2}=\int_0^1 Ur\left\{C_{04}+\frac{C_{05}T}{G}+\frac{C_{06}T^2}{G}\right\}\mathrm{d}r \quad (115.14)$$

$C_{01}=C_1+C_2T'_0-C_3T'^2_0$ $\qquad C_{04}=C_4-C_5T'^2_0+C_6T'^2_0$

$C_{02}=(1/G.C_5)(C_2-2C_3T'_0)$ $\qquad C_{05}=2T'_0(C_6/C_5)-1$

$C_{03}=-C_3/G^2.C_5^2$ $\qquad C_{06}=C_6/C_5^2.G$

The boundary conditions become:

(1) $U=1,\quad V=0,\quad P=P_0,\quad T=T_0;$

(2) $V=0,\quad \frac{\partial U}{\partial r}=0,\quad \frac{\partial T}{\partial r}=0;$

(3) $U=0,\quad V=0;$

(4) (*a*) $\frac{Q}{A}=k\left(\frac{\partial T}{\partial r}\right)_1\left(\frac{\nu_A^2}{R^4C_5g}\right)$ $\qquad$ (*b*) $T=T_R$

COMPUTER PROGRAM

The finite difference equations enumerated in Appendix 115.1 were solved using the following basic procedure.

The temperature equation was written for each radial position at the first axial step from entry, and the (known) coefficients and right-hand sides were evaluated. This gave $M-1$ equations for $M-1$ unknowns (the temperatures), the solution being obtained by matrix inversion.

All the other equations were similarly written, the coefficients evaluated, and matrix inversion used to obtain the unknown velocity and pressure profiles. Where necessary, temperature data from the first inversion were used in the coefficients.

The above procedure was then repeated for the next axial step, the input for it being partly given by output data from the first step. Hence, output data for the entire desired axial distance could be obtained.

The essential features of this approach, therefore, are: (*a*) use of velocities as variables; (*b*) solution of equations at one axial position by matrix inversion; and (*c*) marching-type solution for the unknown field. Various adiabatic laminar flow treatments exist, and as a complete contrast, for example, Friedmann *et al.* (**7**) write the equations in terms of stream function and vorticity, and solve by relaxation over the whole field. As two examples of treatment for turbulent flow, Patankar and Spalding (**8**), and Bankston and McEligot (**9**) use a marching-type procedure, the former solving at one position by successive substitution and the latter by iteration.

The computer program was written in Fortran IV to carry out the above scheme. Programming details with a logical flow diagram are given in reference (**10**). Fairly extensive initial work (**11**) was carried out for adiabatic flows.

Computing facilities for this study were initially provided by The City University ICT 1900 computer. With the number of radial positions fixed at 30, two runs—each of about 40 axial positions—took approximately 50 minutes of computer time.

Subsequent work was carried out using the University of London CDC 6600 computer. This reduced the time taken for solution by a factor of about 30, and permitted more runs and more axial positions to be undertaken simultaneously.

PROPERTY DATA

Property values for air were obtained from reference (**12**), the density–viscosity relationships being calculated for the range 290–490K. The property reference point (subscript A in the notation list) was taken at 290K (i.e. 17°C), which gave a convenient entrance temperature for the heating cases.

Corresponding values for water were obtained (as for reference (**5**)) from references (**13**) and (**14**). The property reference point and entrance temperature were 20°C.

The property coefficients, and other input data requiring treatment, were calculated by hand to six significant figures. It is intended to adapt the input section of the program for future work so that external calculation is reduced to a minimum.

HEATING IN UPFLOW

The main program of work used air as the fluid. Since upward flow with constant wall heat flux has been the case of most interest, it was treated first. For this case the two convections are upward.

To investigate the combined convection processes it is useful to take the case of adiabatic developing laminar flow as a standard of comparison. Results for [illegible] = 0 are plotted in Fig. 115.1; they are taken from reference (**11**) and are valid for all tube orientations.

Temperature affects both the viscosity and the density of the fluid. Initially, three runs were made for the same flow and temperature conditions, viz. (A) variable viscosity (constant density), (B) variable density (constant viscosity), and (C) both properties variable. The heating rate gave a common 20 degC temperature rise.

Effect of viscosity

Data from run (A), corresponding to the adiabatic results, are shown in Fig. 115.1. For air, the viscosity increases with temperature, resulting in a certain accentuation of the developing process. In fact the accepted 'development

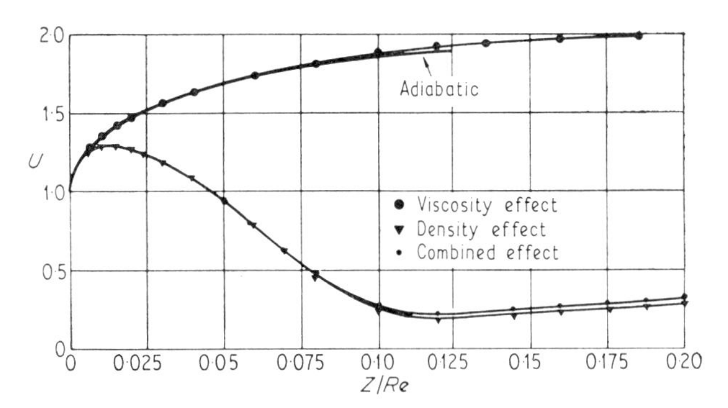

Fig. 115.1. Effect of property changes on centre-line velocity. Constant heat flux at wall—upward flow

length' value of U (viz. 1·98) is reached at $Z/Re = 0{\cdot}186$ instead of the adiabatic 0·25.

Inspection of Fig. 115.1 shows that, relative to density, the effect of viscosity is small and in the opposite sense. It may be concluded, therefore, that the viscosity effect is minor but not negligible.

Effect of density

Results from run (B) are also indicated in Fig. 115.1. For air, the density decreases with temperature, which gives the standard (natural convection) buoyancy force. However, this acceleration of flow initially occurs close to the wall only, since the heating results in fairly high temperatures in this region and negligible temperature changes towards the centre. The acceleration is sufficient to cause the velocity at $r = 0{\cdot}9$ to start increasing after $Z/Re = 0{\cdot}01$. Because of continuity, the centre-line velocity correspondingly starts to decrease, having reached a maximum value of 1·285. This process continues up to about $Z/Re = 0{\cdot}12$, when U at $r = 0$ and 0·9 is 0·186 and 1·017 respectively. After this distance, however, the velocity profile begins to recover, and at $Z/Re = 0{\cdot}20$, U at the centre has reached 0·273.

The cause of this recovery is readily discernible from the computer output, but not easy to show graphically. The normalized temperature profile decreases in range from the entry onwards. At $Z/Re = 0{\cdot}12$, for example, the temperatures (normalized about their 'mixing-cup' mean) range from 0·489 at the centre to 1·459 at $r = 0{\cdot}9$. The moderate temperature range results in less density variation, and hence buoyancy variation, across the duct. The velocity results show that the accelerating part of the velocity field moves steadily towards the centre. At $Z/Re = 0{\cdot}02$ the field between $r = 0{\cdot}7$ and the wall is accelerating, while at $Z/Re = 0{\cdot}112$ the edge of the field has moved to $r = 0{\cdot}5$. At the next axial position the flow adjacent to the wall starts to decelerate (the viscous drag overcomes the buoyancy effect), and at $Z/Re = 0{\cdot}12$ the accelerating part is between $r = 0{\cdot}33$ and 0·8. From then onwards the field accelerates in the inner part of the tube, as it does with adiabatic developing flow.

A third density effect is apparent from the results; the accelerating part of the field progressively widens until from $Z/Re = 0{\cdot}184$ it fills the entire tube. For air, the absolute density level decrease with temperature itself causes a continuing acceleration which exists over the whole field, and is specified by the integral continuity equation. It only shows up uniquely when other forces are approximately in balance.

It may be concluded that changes in density have marked effects on the fluid flow field which lead to distortion and a degree of recovery of the velocity profile, with an overall acceleration of flow.

Combined effect

Results from run (C) are not substantially different from run (B) near the entry, but differences in the velocity profile increase as the viscosity rises with temperature. At $Z/Re = 0{\cdot}2$, for instance, the centre-line U is about 15 per cent higher, and the velocities nearest to the wall ($r = 0{\cdot}933$ outwards) are still decreasing. This is consistent with a relatively increased viscous drag.

Developing temperature and velocity profiles are shown in Figs 115.2 and 115.3 respectively. Pressure profiles are very weak in comparison, and are discussed in detail in reference (11) for adiabatic flows.

Effect of heat flux

Developing centre-line velocities for various heat flux rates are shown in Fig. 115.4. For comparative purposes the terminal temperature changes indicate the heat flux rates.

As the heat flux increases, the minimum velocity reached decreases considerably, even becoming negative at the highest heat flux.

The question arises whether this negative part of the flow-field has any real meaning. While Scheele and Hanratty's results are for water, they did find that a limited amount of reversed flow was stable. In this case the region involved is quite small, and it is postulated, therefore, that it could well be valid.

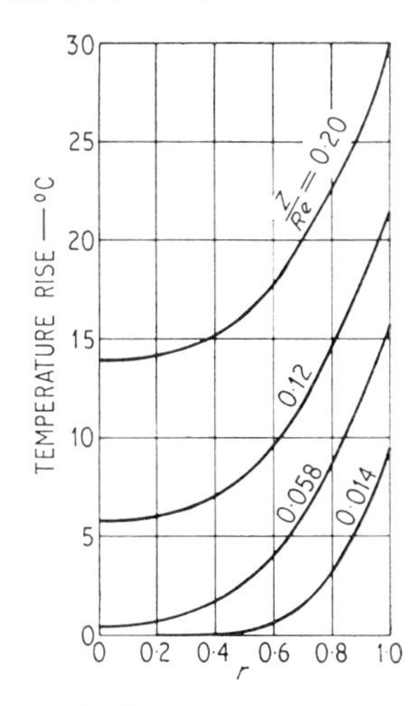

Fig. 115.2. Developing temperature profiles—as for combined effect of Fig. 115.1

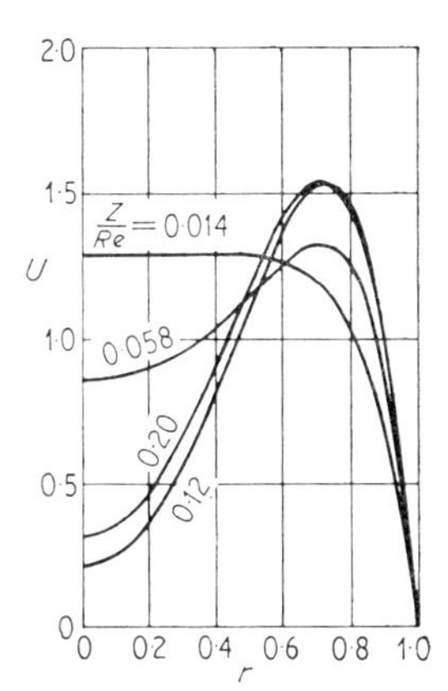

Fig. 115.3. Developing velocity profiles—as for combined effect of Fig. 115.1

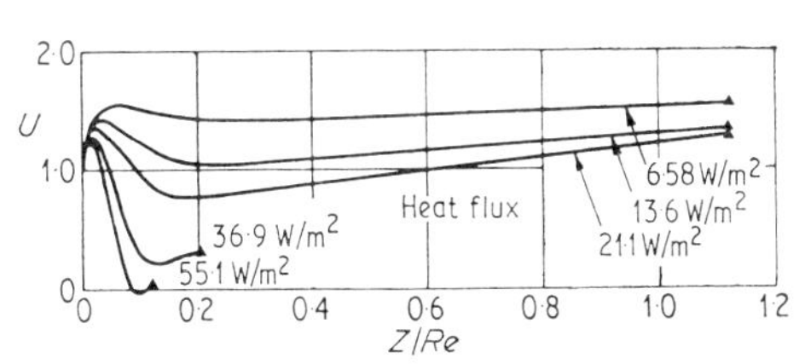

Fig. 115.4. Upward flow. Effect of wall heat flux on centre-line velocity

Two other effects can be noted: first, the distance from entry for this minimum decreases steadily; second, the gradient of the recovery section increases with heat flux. Two factors contribute to this: first, the density decrease with temperature is greater, thus causing a greater relative acceleration; and second, the more severe the velocity distortion with temperature, the more marked will be the recovery effect when viscous effects are no longer suppressed.

HEATING IN DOWNFLOW

Here, the natural convection is upward, in opposition to the forced flow. It has already been shown that the viscosity temperature effect is small, but in this case will act in conjunction with the density effect.

Effect of density and heat flux

Broadly speaking, effects are opposite to those for heating in upflow. The heating initially causes a density decrease close to the wall, which decelerates the flow relative to the adiabatic values. Correspondingly, the centre-line flow is accelerated. If the heating effect is sufficient, the velocity near the wall ($r = 0{\cdot}967$) may become negative (Fig. 115.5). As with upward flow at least a part of this is stable.

In addition, recovery of the profile is just discernible, although in this case the acceleration of the centre-line flow caused by the density decrease with temperature acts against the recovery effect and leads to a more stable profile.

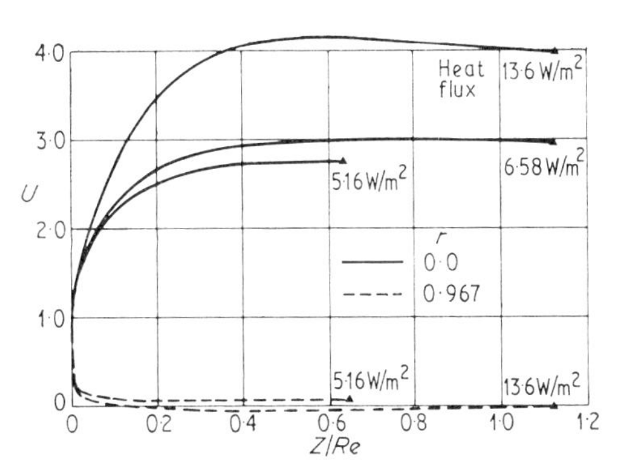

Fig. 115.5. Downward flow. Effect of wall heat flux on velocities at centre and near wall

COOLING CASES FOR AIR

Cooling in downflow and upflow are similar to heating in upflow and downflow in that the natural convection effect respectively assists and opposes the forced convection. Investigation was confined to the nature of this similarity.

It was considered more meaningful to cool from a higher than ambient temperature than from ambient downwards. In fact, the cooling temperature ranges were made identical to their corresponding heating temperature ranges. This had the advantage of eliminating bulk property variation from the comparison.

Redefinition of parameters

Because the outlet cooling temperature was the same as the inlet heating temperature, and the mean temperatures were identical, a redefinition of some parameters and boundary conditions was made.

First, the subscript '0' was used for outlet instead of inlet, thus relating quantities to the same base conditions.

The entry temperature and pressure coincided with the mean exit results for the comparable heating case. The entry axial velocity (uniform profile) was scaled from the heating case value of unity, by the density ratio appropriate to the temperature difference.

Discussion

The main point at issue is whether the cooling cases are significant in their own right or are only negligibly different from the heating ones. Velocity results for the last axial position indicate the former. One reason for this is that the cooling and heating exit temperatures are 40 degC (assisting convection) or 20 degC (opposing convection) different, which means that the relevant densities (and hence the mean velocities) will differ by about 6·5 and 3·3 per cent respectively. This effect has been eliminated in Fig. 115.6 by normalizing the velocity profiles about their mean values.

Fig. 115.6 shows that the profiles are certainly not identical. This can only be because the density (viscosity effects having been shown to be small) for the heating cases starts high and then decreases, but for cooling starts low and then increases over the same range. The profiles,

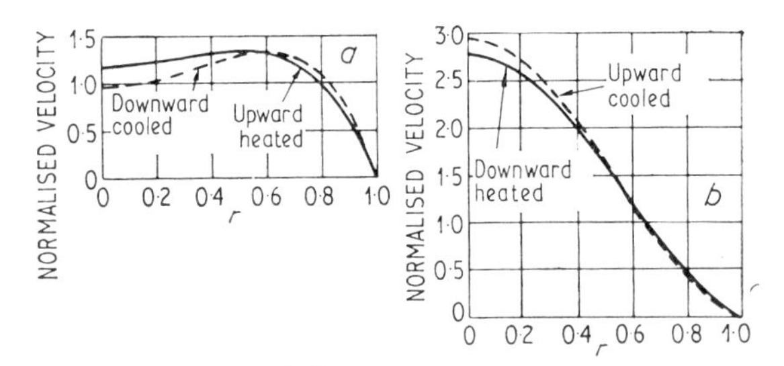

a Assisting.
b Opposing forced convection.

Fig. 115.6. Comparison of normalized velocity profiles for natural convection

incidentally, are closer in absolute and relative terms for opposed convection. This is consistent with a comparison of Figs 115.4 and 115.5, which shows that centre-line velocity changes with distance are substantially less for opposed flow, after initial entry.

If under these restricted comparative conditions the profiles are distinct, then when absolute velocities are compared, and even more so if the cooling temperature range is different from the corresponding heating range (e.g. cooling from ambient), the flow patterns will be quite separate.

This gives a difference between the behaviour of air and water. Scheele and Hanratty (**6**) report the same results for heating in upflow and cooling in downflow. This could be because absolute density changes for water are much less than for air.

CONSTANT WALL TEMPERATURES

This wall boundary condition has one fundamental difference to that of constant heat flux; it admits the possibility of fully developed flow some time after the fluid reaches the wall temperature. This fully developed condition will be that of constant fluid temperature and parabolic velocity profile.

Two main cases were investigated, viz. higher wall temperature than fluid, with flow either upward or downward.

Upward flow

Centre-line velocities are shown in Fig. 115.7 for wall temperatures 15, 20, and 25 degC above ambient. Results may be compared with those of Fig. 115.4, the corresponding heat-flux data. The constant temperature results may be summarized as being sharper in flow field effects, both in reaching velocity minima and in recovery. This is because, although the wall temperatures are fairly modest, the equivalent heat flux near entry is high, decreasing to zero when the fluid temperature has reached that of the wall. For example, the mean fluid temperature (of the 25°C run) is 20°C at $Z/Re = 0{\cdot}15$; i.e. the heat transferred is 80 per cent of that over the total distance of 1·2, where the fluid temperature is almost that of the wall.

This particular temperature point virtually coincides with one of the constant heat flux runs, and the sharper velocity effects, including a negative velocity field, are evident. The possibility of an area of stable reversed flow has already been mentioned as arising from the results of reference (**6**).

Lastly, at $Z/Re = 1{\cdot}2$, the centre-line velocity for the 25°C run is 2·143. When the overall density effect is eliminated by dividing by the mean velocity, the value becomes 1·98, i.e. the accepted development-length condition.

Downward flow

Velocities for the centre and adjacent to the wall ($r = 0{\cdot}967$) are shown in Fig. 115.8. The comments made above for upward flow are also applicable here. Again, a region of reversed flow is predicted for 12·5°C wall temperature and above. Comparing Figs 115.7 and 115.8 it can be said that development lengths for downward flow are shorter.

COMPARISON WITH WATER

The main investigation has been of the behaviour of air, experimental data for which are lacking.

Two comparative runs for water were made for upward and downward heated flow. The same value of non-dimensional heat flux (i.e. Grashof number) and Reynolds number were used as for the corresponding air cases. The centre-line velocities are shown in Fig. 115.9. The effects for water are more marked, and in fact the negative velocity conditions are reached so strongly that the computer results quickly become unstable and meaningless.

There are several reasons for this difference. First, the thermal diffusivity ($= k/\rho C_p$) of water is only about 0·7

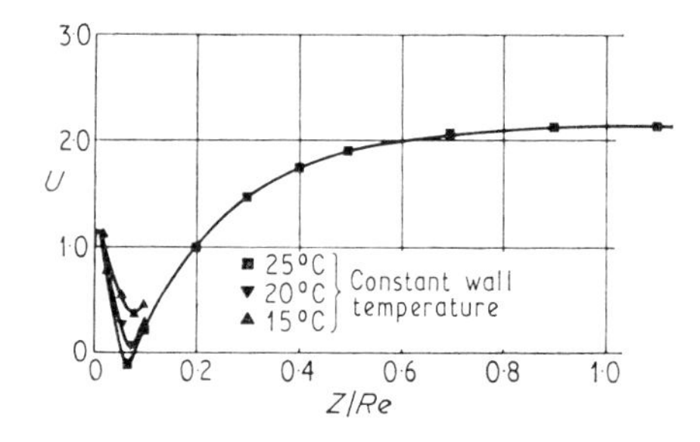

Fig. 115.7. Upward flow. Effect of wall temperature on centre-line velocity

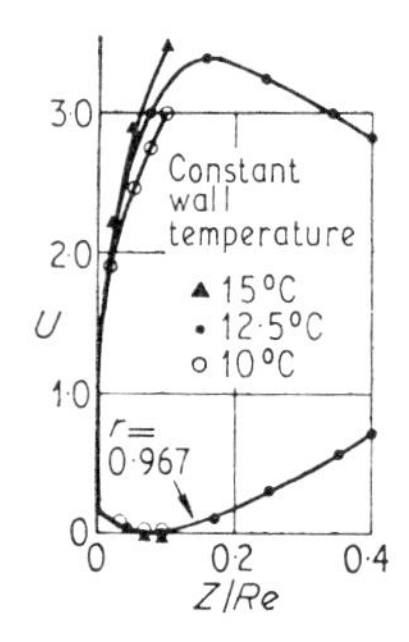

Fig. 115.8. Downward flow. Effect of wall temperature on velocities at centre and near wall

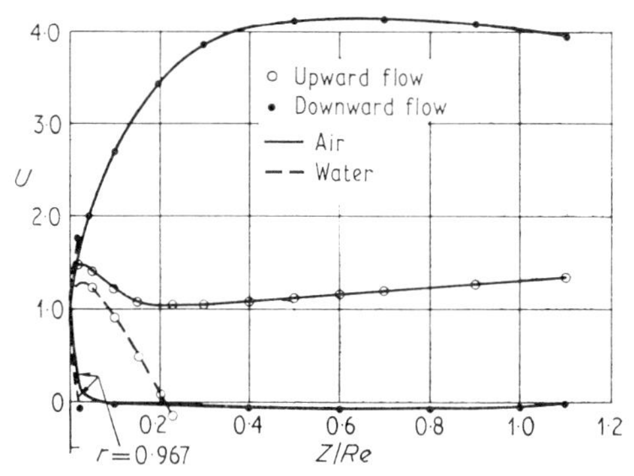

Fig. 115.9. Comparison of behaviour of air and water at same Grashof and Reynolds numbers ($r = 0{\cdot}0$ unless stated)

per cent that of air, which means that only the edges of the flow heat up appreciably, the central portion remaining at ambient for these runs. This accentuates acceleration near the wall. Second, while for air the viscosity effect slightly reduces the density effect, for water the viscosity sharply reduces with temperature, thus substantially aiding the density effect. Last, the absolute change of density with water is so small it may be neglected. This removes the bulk acceleration of flow that has been noted with air.

Therefore, from this brief summary it is demonstrated that the mixed convection effects, from every cause, are stronger for water than air.

TRANSITION

Transition to a form of turbulence is discussed at some length in references (**4**) and (**5**). The overall conclusion is that this is associated with the velocity profile shape, with only a small dependence on Reynolds number. For upward heated flow the profile is unstable if the central velocity is less than that at some radius, although this instability takes some axial distance to develop. For downward heated flow, transition takes place very quickly after a negative velocity is reached at the wall, and this is the reason for emphasizing this condition, where appropriate, in the present study. The transition criterion can only be found empirically. One reason for including the radial momentum equation in this work was to find whether any radial pressure variation exists when instability occurs, but in fact the pressure profiles are always uniform. An immediate conclusion is that corresponding transition data are needed for air.

CONVECTIVE HEAT TRANSFER

This work has concentrated on the basic flow processes involved in the various combined convection cases. It should be noted also that corresponding thermal cases, with flow in either direction, give significantly different heat transfer results as distance increases. Two examples are given.

Where the heat flux is constant (viz. temperature change of 40 degC over $Z/Re = 1{\cdot}12$), Nusselt numbers and wall temperatures are different. At the distance quoted, for upflow the quantities are 1·88 times and 0·905 times the respective downflow values, i.e. implying better wall cooling for upflow.

Where the wall temperature is constant (viz. 15°C) the Nusselt number and mean fluid temperature for upflow at $Z/Re = 0{\cdot}1$ are 2·18 and 1·35 times those for downflow, implying substantially better convection for upflow.

Hence, as would be expected, where the free convection assists the forced, the heat transfer is improved.

ACCURACY

The reliability and accuracy of the program were checked in two ways: against adiabatic and non-adiabatic data.

By removing any temperature effects, the conditions became those of adiabatic developing laminar flow. This was quite fully investigated (**11**) and the comparison proved satisfactory. The only limitation was that of instability, which determined a minimum level to the axial distance increment at entry. This depended on Reynolds number.

An attempt was then made to predict a set of Lawrence and Chato's published results. Difficulty was encountered in calculating the appropriate input data from that available, but again the comparison was good. The reservation in this case was that the choice of changes in axial distance increment (Δz) had an effect on the velocities. However, such effects tended to be local.

The above gave confidence that, used with care, the program would predict reliable results. In the constant heat flux case, a further check was that of a cumulative heat balance, and in general a difference of 1 per cent was attained. For a constant wall temperature this check was not possible, but the Δz pattern was chosen to correspond with the most comparable heat flux case.

The final consideration is that of validity of treatment for partially reversed flows. While the concept of reversal and recovery have experimental and logical validity, some reservation must be expressed as to detailed accuracy. First, the step-by-step solution is intended for an upstream progression, whereas for reversed flow the sequence would be downstream. Second, especially where reversal is at the wall, the temperature of the reversed fluid should increase in the backward direction, contrary to the prediction. Where reversed effects are small, this may not be too serious. In any case, for strong effects, instability occurs in the computer treatment.

FUTURE WORK

As it stands, the program may be used in two ways. The comparative study made here may be made more comprehensive and detailed. In addition, data may be obtained from a range of relevant fluids.

The cylindrical flow field studied here is the same for an annulus, except for the inner boundary conditions. The flexibility of the program enables it to be partially rewritten in a form suitable for annular flow. This is being considered (**15**).

There is some indication (16) that by incorporating eddy and thermal diffusivities it may be possible to treat turbulent flow with the same basic equations. The necessary choice of a turbulence model (see reference (17)) requires considerable care. While ease of treatment lends bias to a simple model, the very combined convection effects looked for may not be predicted.

CONCLUSIONS

The controlling partial differential equations for heat transfer by laminar combined convection in a vertical tube, have been re-expressed in finite difference form and solved by means of a digital computer program in Fortran IV.

Results confirm the need for a developing (two-dimensional) flow model to describe this problem in different cases.

The effect of the major parameters has been studied on a comparative basis, and interpretation of results has been aided by the detailed data for the velocity and temperature fields.

Transition to a form of turbulence can only be predicted using experimental results; while some exist for water, there is a need for corresponding data for gases. Future developments include more detailed investigation, flow in vertical annuli, and the consideration of a turbulent flow model.

APPENDIX 115.1

FINITE DIFFERENCE EQUATIONS

The dimensionless equations are re-expressed using standard finite difference techniques, as applied by Lawrence and Chato (5). The dimensionless radius r is re-expressed as $r=(m-1)/M$, for $m=1,2,\ldots,M+1$, i.e. for $m=1$, $r=0$ (axis), and for $m=M+1$, $r=1$ (wall). This gives a rectangular (axial and radial) mesh with suffices n and m respectively. Equations (115.8)–(115.11) and (115.14) become (115.15)–(115.19) respectively, written in linear form with suffix $n+1$ representing the unknown axial position to be solved. The (known) constants relevant to each variable are in square brackets, as are the 'known' right-hand sides of equations.

$$\left[\frac{2Mm\,\Delta z}{m-1}\right]V_{n+1,m}-[2\Delta zM]V_{n+1,m-1}+[1]U_{n+1,m}+[1]U_{n+1,m-1}=[U_{n,m}+U_{n,m-1}] \quad (115.15)$$

$$\left[\frac{M^2}{2Re}\left\{T_A\frac{(2m-1)}{m-1}+\frac{T_B}{2}(T_{n+1,m+1}-T_{n+1,m-1})\right\}-\frac{M}{2}V_{n,m}\right]U_{n+1,m+1}+\left[\frac{1}{Re}\left\{T_A\left(\frac{1}{\Delta z^2}-2M^2\right)+\frac{2T_B}{\Delta z^2}(T_{n+1,m}-T_{n,m})\right\}-\frac{U_{n,m}}{\Delta z}\right]U_{n+1,m}+\left[\frac{M^2}{2Re}\times\left\{T_A\frac{(2m-3)}{(m-1)}-\frac{T_B}{2}(T_{n+1,m+1}-T_{n+1,m-1})\right\}+\frac{M}{2}V_{n,m}\right]U_{n+1,m-1}-\frac{P_{n+1,m}}{[\Delta z]}$$

$$=\left[\pm\frac{1}{Re^2}\left\{G.C_{04}+T_{n+1,m}(C_{05}+C_{06}T_{n+1,m})\right\}-\frac{P_{n,m}}{\Delta z}-\frac{U_{n,m}{}^2}{\Delta z}+\frac{2U_{n,m}}{\Delta z^2.Re}\{T_A+T_B(T_{n+1,m}-T_{n,m})\}-\frac{U_{n-1,m}}{\Delta z^2.Re}T_A-\frac{MT_B}{2Re\,\Delta z}(T_{n+1,m+1}-T_{n+1,m-1})\times(V_{n,m}-V_{n-1,m})\right] \quad (115.16)$$

$$\left[\frac{M^2}{2Re}\left\{\frac{T_A(2m-1)}{(m-1)}+T_B(T_{n+1,m+1}-T_{n+1,m-1})\right\}-\frac{M}{2}V_{n,m}\right]\times V_{n+1,m+1}+\left[\frac{1}{Re}\left\{T_A\left(\frac{1}{\Delta z^2}-\frac{M^2}{(m-1)^2}(2m^2-4m+3)\right)+\frac{T_B}{\Delta z^2}(T_{n+1,m}-T_{n,m})\right\}-\frac{U_{n,m}}{\Delta z}\right]V_{n+1,m}+\left[\frac{M^2}{2Re}\left\{T_A\frac{(2m-3)}{(m-1)}-T_B(T_{n+1,m+1}-T_{n+1,m-1})\right\}+\frac{M}{2}V_{n,m}\right]V_{n+1,m-1}+\left[\frac{M}{12}\right]P_{n+1,m+2}-\left[\frac{2M}{3}\right]P_{n+1,m+1}+\left[\frac{2M}{3}\right]P_{n+1,m-1}-\left[\frac{M}{12}\right]P_{n+1,m-2}$$

$$=\left[-\frac{U_{n,m}.V_{n,m}}{\Delta z}+\frac{2V_{n,m}.T_A}{Re.\Delta z^2}+\frac{V_{n-1,m}}{\Delta z^2.Re}\left\{\frac{T_B}{2}(T_{n+1,m}-T_{n,m})-T_A\right\}-\frac{MT_B}{2Re\,\Delta z}(T_{n+1,m}-T_{n,m})\times(U_{n,m+1}-U_{n,m-1})\right] \quad (115.17)$$

$$T_A=[C_{01}+T_{n+1,m}(C_{02}+C_{03}T_{n+1,m})]$$

$$T_B=(C_{02}+2C_{03}T_{n+1,m})$$

$$\left[\frac{M}{2}V_{n,m}+\frac{M^2(1-2m)}{2Pr.Re(m-1)}\right]T_{n+1,m+1}+\left[\frac{U_{n,m}}{\Delta z}+\frac{1}{Pr.Re}\left(2M^2-\frac{1}{\Delta z^2}\right)\right]T_{n+1,m}+\left[\frac{M^2(3-2m)}{2Pr.Re(m-1)}-\frac{M}{2}V_{n,m}\right]T_{n+1,m-1}$$

$$=\left[\left\{U_{n,m}-\frac{2}{Pr.Re\,\Delta z}\right\}\frac{T_{n,m}}{\Delta z}+\frac{T_{n-1,m}}{\Delta z^2.Pr.Re}\right] \quad (115.18)$$

$$\frac{3M^2.C_{04}}{8}=\sum_{m=1}^{M/2}\left[(2m-1)U_{n+1,2m}\left\{C_{04}+\frac{C_{05}}{G}T_{n+1,2m}+\frac{C_{06}}{G}T^2_{n+1,2m}\right\}\right]+\sum_{m=1}^{M/2-1}\left[m.U_{n+1,2m+1}\left\{C_{04}+\frac{C_{05}}{G}T_{n+1,2m+1}+\frac{C_{06}}{G}T^2_{n+1,2m+1}\right\}\right] \quad (115.19)$$

Equation (115.19) is obtained by applying Simpson's rule to the integral continuity equation. The boundary conditions become:

(1) $U_{1,m} = 1, \quad V_{1,m} = 0, \quad P_{1,m} = P_0, \quad T_{1,m} = 0$

(2) $U_{n,1} = U_{n,2}, \quad V_{n,1} = 0, \quad P_{n,1} = P_{n,2},$
$T_{n,1} = T_{n,2}$

(3) $U_{n,M+1} = 0, \quad V_{n,M+1} = 0$

(4) (a) $T_{n,M+1} = T_{n,M} + \dfrac{QR^4C_5g}{A\nu_A{}^2Mk}$; (b) $T_{n,M+1} = T_R$

APPENDIX 115.2

REFERENCES

(1) COLLINS, M. W. 'Mixed free and forced convection—a survey', The City University, Dept Mech. Engng, Research Memo. No. ML15, 1969.

(2) MARTINELLI, R. C. and BOELTER, L. M. K. 'The analytical prediction of superposed free and forced viscous convection in a vertical pipe', *Univ. Calif. Publs Engng* 1942 **5** (2), 23.

(3) IQBAL, M. and STACHIEWICZ, J. W. 'Influence of tube orientation on combined free and forced laminar convection heat transfer', *J. Heat Transfer, Trans. Am. Soc. mech. Engrs* 1966, 109.

(4) ROSEN, E. M. and HANRATTY, T. J. 'Use of boundary-layer theory to predict the effect of heat transfer on the laminar-flow field in a vertical tube with a constant temperature wall', *A.I.Ch.E. Jl* 1961 **7** (No. 1), 112.

(5) LAWRENCE, W. T. and CHATO, J. C. 'Heat-transfer effects on the developing laminar flow inside vertical tubes', *J. Heat Transfer, Trans. Am. Soc. mech. Engrs* 1965, Paper No. 65-WA/HT-11.

(6) SCHEELE, G. F. and HANRATTY, T. J. 'Effect of natural convection on stability of flow in a vertical pipe', *J. Fluid Mech.* 1962 **14**, 244.

(7) FRIEDMANN, M., GILLIS, J. and LIRON, M. 'Laminar flow in a pipe at low and moderate Reynolds numbers', *Appl. scient. Res.* 1968 **19**, 426.

(8) PATANKAR, S. V. and SPALDING, D. B. *Heat and mass transfer in boundary layers* 2nd edit., 1970.

(9) BANKSTON, C. A. and MCELIGOT, D. M. 'Turbulent and laminar heat transfer to gases with varying properties in the entry region of circular ducts', *Int. J. Heat Mass Transfer* 1970 **13**, 319.

(10) COLLINS, M. W. 'Combined convection in vertical tubes—a computer program in Fortran IV', The City University, Dept Mech. Engng, Research Memo. No. ML25, 1970.

(11) COLLINS, M. W. 'Developing laminar flow in a circular tube', University of Salford Symposium on Internal Flows, 1971.

(12) 'Tables of thermal properties of gases', *Nat. Bur. Stand. Circ. 564* 1955.

(13) KEMENY, G. A. and SOMERS, E. V. 'Combined free and forced-convective flow in vertical circular tubes—experiments with water and oil', *J. Heat Transfer, Trans. Am. Soc. mech. Engrs* 1962 **84** (Series C).

(14) *Handbook of chemistry and physics* 38th edit., 1956, 1989 (Chemical Rubber Publishing Co., Cleveland).

(15) COLLINS, M. W. 'Combined convection—further computer developments, and a program for annuli', The City University, Dept Mech. Engng, Research Memo No. ML28, 1971 (in preparation).

(16) OJALVO, M. S., ANAND, D. K. and DUNBAR, R. P. 'Combined forced and free turbulent convection in a vertical circular tube with volume heat sources and constant wall heat addition', *J. Heat Transfer, Trans. Am. Soc. mech. Engrs* 1967, 328.

(17) LAUNDER, B. E. and SPALDING, D. B. 'Turbulence models and their application to the prediction of internal flows', University of Salford Symposium on Internal Flows, 1971.

C116/71 COMBINED FREE AND FORCED LAMINAR CONVECTION IN THE ENTRANCE REGION OF DUCTS OF CONSTANT CROSS-SECTION

T. L. SITHARAMARAO* H. BARROW†

An approximate analysis of the effects of free convection on a predominantly forced convection laminar flow in the entrance region of a vertical parallel wall channel is presented. The theoretical model is a flat vertical plate at a constant temperature, immersed in an infinte medium and subject to a varying free stream velocity parallel to the plate. Similar solutions which are valid for small values of the controlling parameter Gr_x/Re_x^2 are obtained up to a first-order approximation. Numerical results have been obtained for fluids with Prandtl numbers equal to 0·72, 1·0, and 5·0 and representative velocity and temperature profiles are given.

INTRODUCTION

IN MANY OF THE so-called forced convection situations encountered in the engineering field, free convection is an essential mechanism in the process and may be sufficiently large to have a significant effect on the fluid flow and heat transfer characteristics.

In recent years much theoretical work has been published on the interaction of free convection on forced convection, and the corresponding case of forced convection effects on a predominantly free convection process (1)–(3)‡. These earlier analyses pertained to the flow of a fluid over a vertical flat plate with uniform stream velocity, although Sparrow *et al.* (4) obtained similar type solutions when the free stream velocity and the surface temperature varied as x^m and x^{2m-1}. Acrivos (5) analysed the interacting free and forced convection about a flat vertical plate by means of the Karman–Pohlhausen method. It has been suggested that the solutions of these earlier studies would also be applicable in the entrance region of ducted flows.

The main difference between the isolated flat plate and the duct problem is that in the latter case the main stream velocity is non-uniform on account of mass continuity requirements. The application of the results of the previously mentioned analyses for mixed convection in a ducted flow will be very approximate (the analysis of reference (4) is exact in itself but situations in which the velocity and temperature are specified are not common).

In the present paper, the effects of free convection on a predominantly forced convection flow over a constant temperature flat plate in a stream with velocity variation $u_\infty \simeq u_0 x^m$ are examined. The chosen value of m is that pertaining to the potential core flow in the region of the entry of a parallel wall channel or, in calculation, a close approximation to it. The potential core velocity must of necessity change to satisfy continuity requirements, and ideally this should be checked.

The MS. of this paper was received at the Institution on 31st March 1971 and accepted for publication on 7th May 1971. 23

* *Birla Institute of Technology and Science, Pilani, India.*
† *University of Liverpool.*

‡ *References are given in Appendix 116.1.*

Notation

c_f	Friction coefficient [$= \tau_w/\frac{1}{2}\rho u_\infty^2$].
c_p	Specific heat at constant pressure.
d_e	Equivalent diameter of parallel walled duct.
F	Dimensionless stream function.
Gr_x	Grashof number [$= g\beta(T_w - T_\infty)x^3/\nu^2$].
g	Acceleration due to gravity.
h	Convective heat transfer coefficient.
K	Thermal conductivity of fluid.
m	Index in equation (116.1).
Nu	Nusselt number [$= hx/K$].
Pr	Prandtl number.
p	Pressure.
q	Heat rate per unit area.
Re	Reynolds number for ducted flow [$= u_0 d_e/\nu$].
Re_x	Reynolds number [$= u_\infty x/\nu$].
T	Temperature.
u	Velocity parallel to plate or axis of duct.
v	Velocity normal to plate or axis of duct.
x	Distance from entry in flow direction.
x^+	Dimensionless distance [$= x/d_e Re$].
y	Distance normal to plate or wall of duct.
α	Thermal diffusivity [$= K/c_p\rho$].

β Coefficient of volume expansion.
θ Dimensionless temperature $[=(T-T_\infty)/(T_w-T_\infty)]$.
λ Blasius similarity variable $[=y(u_\infty/\nu_x)^{1/2}]$.
μ Absolute viscosity of fluid.
ν Kinematic viscosity of fluid $[=\mu/\rho]$.
ρ Density of fluid.
τ Shear stress.
ψ Stream function, equation (116.10).

Suffixes

fc Pure forced convection value.
w Wall.
0 Condition at $x = 0$.
∞ Free stream.

DERIVATIONS OF THE EQUATIONS

Consider the upward flow of a fluid at temperature T_∞ over a heated semi-infinite flat vertical plate. The plate is maintained at temperature T_w and the flow is influenced by a gravitational body force field. The fluid motion is considered to be steady and laminar throughout, and the free stream velocity at relatively large distance from the wall is expressed by the relation

$$u_\infty \propto u_0 x^m \quad . \quad . \quad . \quad . \quad (116.1)$$

where u_0 (the free stream velocity at $x = 0$) and m are constants. This type of variation of free stream velocity is met in 'wedge flows' and in the entry region of ducts.

All fluid properties, with the exception of the density, in the body force term of the momentum equation are considered constant. Following accepted practice, the density (ρ) is taken to be a linear function of temperature through the coefficient of volume expansion (β).

The boundary layer forms of the mass, momentum, and energy equations for the non-dissipative flow are (**6**):

$$\frac{\partial u}{\partial x}+\frac{\partial v}{\partial y}=0 \quad . \quad . \quad . \quad . \quad (116.2)$$

$$\rho\left(u\frac{\partial u}{\partial x}+v\frac{\partial u}{\partial y}\right)=-\frac{\partial p}{\partial x}-\rho g+\mu\frac{\partial^2 u}{\partial y^2} \quad (116.3)$$

and

$$u\frac{\partial T}{\partial x}+v\frac{\partial T}{\partial y}=\alpha\frac{\partial^2 T}{\partial y^2} \quad . \quad . \quad (116.4)$$

(The term $\alpha(\partial^2 T/\partial x^2)$ is omitted from equation (116.4) on the understanding that Pe, the Peclet number, of the flow is sufficiently large when axial conduction is negligible.)

The boundary conditions are

$$\left.\begin{array}{lll} u=v=0, & T=T_w, & y=0 \\ u=u_\infty, & T=T_\infty, & y\to\infty \end{array}\right\} \quad (116.5)$$

The pressure gradient term in equation (116.3) can be expressed in terms of the free stream velocity through Bernoulli's equation

$$\frac{\mathrm{d}p}{\mathrm{d}x}+\rho_\infty u_\infty\frac{\mathrm{d}u_\infty}{\mathrm{d}x}+\rho_\infty g=0 \quad . \quad (116.6)$$

which is valid for the potential flow region.

Substituting equation (116.6) in equation (116.3), and with

$$\rho=\rho_\infty[1-\beta(T-T_\infty)] \quad . \quad . \quad (116.7)$$

we obtain

$$u\frac{\partial u}{\partial x}+v\frac{\partial u}{\partial y}=u_\infty\frac{\mathrm{d}u_\infty}{\mathrm{d}x}+g\beta(T-T_\infty)+\nu\frac{\partial^2 u}{\partial y^2} \quad (116.8)$$

where it is tacitly assumed that $\rho_\infty/\rho \simeq 1$ as is accepted practice.

From equation (116.1),

$$\frac{\mathrm{d}u_\infty}{\mathrm{d}x}=\frac{mu_\infty}{x} \quad . \quad . \quad . \quad (116.9)$$

and introducing the stream function, ψ, where

$$u=\frac{\partial\psi}{\partial y} \quad \text{and} \quad v=-\frac{\partial\psi}{\partial x} \quad . \quad (116.10)$$

equation (116.8) becomes

$$\frac{\partial\psi}{\partial y}\cdot\frac{\partial^2\psi}{\partial x\,\partial y}-\frac{\partial\psi}{\partial x}\cdot\frac{\partial^2\psi}{\partial y^2}=\frac{mu_\infty{}^2}{x}+g\beta(T-T_\infty)+\nu\frac{\partial^3\psi}{\partial y^3} \quad . \quad . \quad . \quad (116.11)$$

The energy equation [equation (116.4)] can also be expressed in terms of the stream function, ψ, and a dimensionless temperature, θ, as

$$\frac{\partial\psi}{\partial y}\cdot\frac{\partial\theta}{\partial x}-\frac{\partial\theta}{\partial y}\cdot\frac{\partial\psi}{\partial x}=\alpha\frac{\partial^2\theta}{\partial y^2} \quad . \quad (116.12)$$

The boundary conditions for the problem of solving equations (116.11) and (116.12) are, in terms of the new variables,

$$\left.\begin{array}{lll} \dfrac{\partial\psi}{\partial y}=\dfrac{\partial\psi}{\partial x}=0, & \theta=1, & y=0 \\ \dfrac{\partial\psi}{\partial y}=u_\infty, & \theta=0, & y\to\infty \end{array}\right\} \quad (116.13)$$

SOLUTION OF THE EQUATIONS

In as much as the analytical solutions of equations (116.11) and (116.12), subject to the boundary conditions (116.13), could not be found, a similarity solution using the perturbation technique is attempted. As the problem is considered to be principally forced convection with superimposed free convection effects, a solution in the form of a series whose first term (zeroth approximation) corresponds to the solution of the pure forced convection case is assumed. In this case the perturbation parameter involves the ratio $Gr_x/Re_x{}^2$ following references (**1**) and (**7**): this ratio measures the relative magnitudes of the buoyancy and the inertia forces. Thus, the solution for the stream function, ψ, and the dimensionless temperature, θ, is assumed in terms of this parameter to be

$$\psi=(\nu x u_\infty)^{1/2}\left[F_0(\lambda)\pm\frac{Gr_x}{Re_x{}^2}F_1(\lambda)\pm\cdots\right] \quad (116.14)$$

and

$$\theta=\frac{T-T_\infty}{T_w-T_\infty}=\left[\theta_0(\lambda)\pm\frac{Gr_x}{Re_x{}^2}\theta_1(\lambda)\pm\cdots\right] \quad (116.15)$$

where λ, Gr_x, and Re_x are, respectively, the Blasius similarity variable, the Grashof number, and the Reynolds number. The plus–minus sign corresponds, respectively, to the case when buoyancy aids the forced flow (heated upward flow) and when buoyancy opposes the forced convection (heated downward flow). From the definition of the stream function, the local velocity, u, may be written

$$u = u_\infty \left[F'_0(\lambda) \pm \frac{Gr_x}{Re_x^2} F'_1(\lambda) \right] \quad (116.16)$$

where the primes denote differentiation with respect to λ and the expression is terminated at the first-order approximation.

On substituting equations (116.14) and (116.15) into equations (116.11) and (116.12), and equating the coefficients of like powers of Gr_x/Re_x^2, we obtain a set of ordinary differential equations for the F's and the θ's up to the first-order approximation as follows:

$$F'''_0 + \left(\frac{m+1}{2}\right) F_0 F''_0 + m(1 - F'^2_0) = 0 \quad (116.17)$$

$$\theta''_0 + \left(\frac{m+1}{2}\right) Pr . \theta'_0 F_0 = 0 \ . \quad (116.18)$$

$$F'''_1 + \left(\frac{m+1}{2}\right) F''_1 F_0 - F'_1 F'_0 + \frac{3(1-m)}{2} F_1 F''_0 + \theta_0 = 0 \quad . \ . \ . \ (116.19)$$

$$\theta''_1 + Pr \left[\left(\frac{m+1}{2}\right) \theta'_1 F_0 - (1-2m)\theta_1 F'_0 + \tfrac{3}{2}(1-m) F_1 \theta'_0 \right] = 0 \quad (116.20)$$

The boundary conditions are now

$$\left.\begin{array}{ll} F_0(0) = F_1(0) = 0 & \\ F'_0(0) = F'_1(0) = 0 & \\ \theta_0(0) = 1, & \theta_0(\infty) = 0 \\ F'_0(\infty) = 1, & F'_1(\infty) = 0 \\ \theta_1(0) = 0, & \theta_1(\infty) = 0 \end{array}\right\} \ . \quad (116.21)$$

Equation (116.17) for the zeroth approximation was solved by Falkner and Skan (**8**) and the results are tabulated for various values of m while the solution to equation (116.18) has been given by Eckert (**9**). These equations must, however, be resolved in order to obtain numerical solutions to equations (116.19) and (116.20).

The Kutta–Merson integration procedure for solving a set of simultaneous differential equations was used. As there are two third-order differential equations and two second-order ones, 10 boundary conditions are required at the starting point $\lambda = 0$. Only six boundary conditions are known at $\lambda = 0$, those remaining referring to conditions at $\lambda = \infty$. Hence the problem is a 'two point' boundary problem, which involves guessing the values of $F''_0(0)$, $F''_1(0)$, $\theta'_0(0)$, and $\theta'_1(0)$, or the 'gradients' of the velocity and temperature profiles at the wall. The integration is carried out up to $\lambda = \infty$ (a value of $\lambda = 9$ was found to be sufficient here) and the results are then compared with the specified values at $\lambda = \infty$. By repeated iteration values with any prescribed accuracy can be achieved.

CALCULATION OF HEAT TRANSFER AND SHEARING STRESS

The dimensionless heat transfer coefficient, or Nusselt number is, in the usual notation,

$$Nu = \frac{hx}{K} = -\frac{x(\partial T/\partial y)_{y=0}}{(T_w - T_\infty)} \ . \quad (116.22)$$

Substituting for $(\partial T/\partial y)$ from equation (116.15):

$$Nu = -Re_x^{1/2} \left[\theta'_0(0) \pm \frac{Gr_x}{Re_x^2} \theta'_1(0) \right] \quad (116.23)$$

If the pure forced convection solution (i.e. the first term on the right-hand side of equation (116.23)) is written as Nu_{fc}, then

$$\frac{Nu}{Nu_{fc}} = 1 \pm \left(\frac{Gr_x}{Re_x^2}\right)\left(\frac{\theta'_1(0)}{\theta'_0(0)}\right) \ . \quad (116.24)$$

where the effect of buoyancy on heat transfer is measured by the value of

$$(Gr_x/Re_x^2)\{[\theta'_1(0)]/[\theta'_0(0)]\}$$

Likewise it is easily shown that

$$\frac{c_f}{(c_f)_{fc}} = 1 \pm \left(\frac{Gr_x}{Re_x^2}\right)\left(\frac{F''_1(0)}{F''_0(0)}\right) \ . \quad (116.25)$$

where the last term measures the influence of the free convection on the friction coefficient, c_f.

RESULTS AND DISCUSSION OF NUMERICAL CALCULATIONS

The numerical solution to equations (116.17)–(116.20) were obtained on an English Electric KDF 9 digital computer for values of $Pr = 0{\cdot}72$, $1{\cdot}0$, and $5{\cdot}0$ for a value of $m = 0{\cdot}0878$. The calculations were carried out to six places of decimals and the error was $< 10^{-5}$. The velocity and temperature functions are shown in Fig. 116.1 for $Pr = 0{\cdot}72$. The corresponding profiles for $Pr = 0{\cdot}72$ and for values of the parameter Gr_x/Re_x^2 equal to 0·1 and 0·2 for both aiding and opposing flows are presented in Fig. 116.2. In Fig. 116.2 the velocity and temperature distributions for the case of constant free stream velocity ($m = 0$) which corresponds to the classical Blasius solution are also indicated. We note that the allowance for variation of the mainstream is to increase the wall shear stress and heat transfer. Furthermore, with aiding flow (i.e. positive values of Gr_x/Re_x^2) there is a further increase in heat transfer, while with opposing flow there is a reduction. Similar graphs for $Pr = 5{\cdot}0$ are shown in Fig. 16.3.

The values of $F''_0(0)$, $\theta'_0(0)$, $F''_1(0)$, and $\theta'_1(0)$ at $\lambda = 0$ for various values of the Prandtl number are shown separately in Table 116.1 to facilitate the calculation of the Nusselt number and the friction coefficient according to equations (116.24) and (116.25). It is to be noted that if the deviation in Nusselt number is limited to 5 per cent, then the upper limit of Gr_x/Re_x^2 is about 0·1. For this

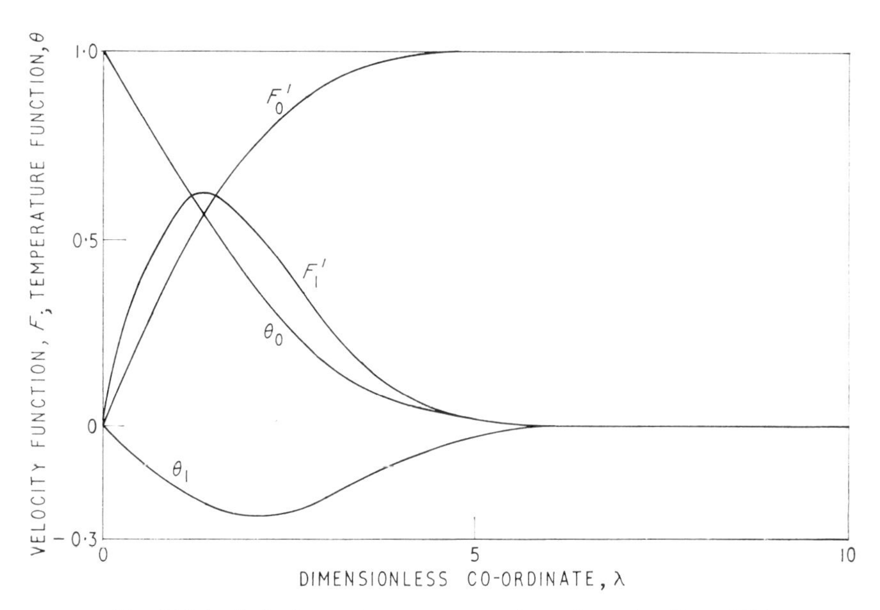

Fig. 116.1. Velocity and temperature functions for $Pr = 0·72$

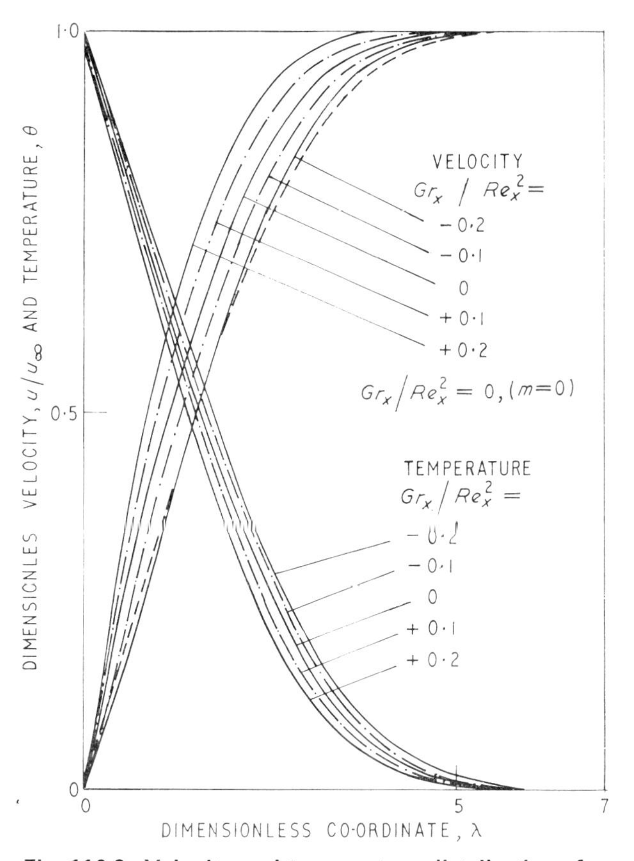

Fig. 116.2. Velocity and temperature distributions for $Pr = 0·72$

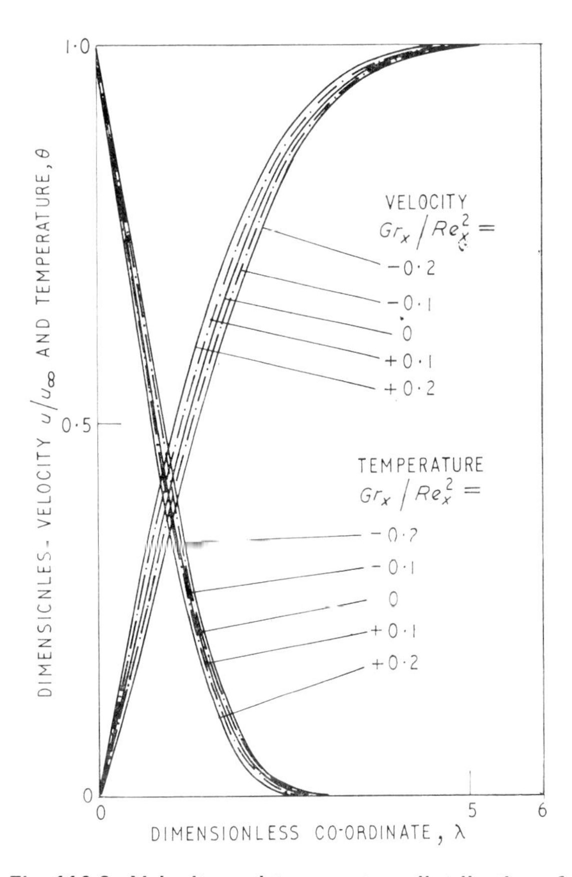

Fig. 116.3. Velocity and temperature distributions for $Pr = 5$

Table 116.1. Values of derivatives of dimensionless stream function and temperature

$m = 0{\cdot}0878$

Pr	$F''_0(0)$	$F''_1(0)$	$\theta'_0(0)$	$\theta'_1(0)$	$\theta'_1(0)/\theta'_0(0)$
0·72	0·479 29	1·019 94	−0·327 716	−0·168 2	0·511 7
1·0	0·479 29	0·961 245	−0·369 59	−0·182 66	0·494 3
5·0	0·479 29	0·679 328	−0·652 893	−0·244 91	0·375 2

value of Gr_x/Re_x^2 the maximum change in the friction coefficient is about 20 per cent.

Although this analysis presents information on the fluid flow and heat transfer characteristics for mixed convection in the entry region of a two-dimensional flow, it cannot predict the region of applicability of the results. The validity of the solutions themselves are controlled by the value of the parameter Gr_x/Re_x^2 for the series expansions to converge. For larger values of this parameter the solutions have to be found to a better degree of approximation.

In order to apply the analysis it is necessary to know the velocity variation of the free stream, i.e. the index m of equation (116.1). The velocity of the free stream very near the entry will not be very different from the pure forced convection value, because the influence of the free convection effects is confined to the boundary layers in that region. Of necessity, m must be different from that in pure forced convection to accommodate the mass continuity requirement. However, as a first approximation, the value for m in pure forced convection suffices, and in the present analysis we have used that value describing the potential core velocity variation for flow between parallel plates (**10**), i.e.

$$m = 0{\cdot}0878[10^{-4} \leqslant x^+ \leqslant 10^{-2}] \qquad (116.26)$$

Further refinement to the analysis would include a variable exponent m to cater for the free convection influence on the main stream, but this would present major computational difficulties and has not been pursued.

In their present form, the analysis and the results can be used to establish the conditions when free convection effects in predominantly forced convection in the present geometry must be taken account of. The value of the parameter Gr_x/Re_x^2 can be assessed and then the percentage increase (or decrease) in the friction and heat transfer coefficients can be evaluated from equations (116.24) and (116.25). We note that because $\theta'_1(0)/\theta'_0(0)$ for water ($Pr \simeq 5$, say) is less than $\theta'_1(0)/\theta'_0(0)$ for air ($Pr \simeq 0{\cdot}72$, say), then for a given value of Nu/Nu_{fc} the parameter Gr_x/Re_x^2 is larger in the case of water. Because the parameter measures the buoyancy effects relative to inertia effects, this last observation is consistent with the fact that water is a better free convector than air.

There are many familiar applications where these latter engineering results are of interest. During shutdown of a nuclear reactor, for example, the coolant velocity through the vertical passages may be sufficiently reduced for free convection to play an increasingly important role in the transfer processes. The ideas may possibly be extended to other body force situations associated with duct rotation and space flight.

APPENDIX 116.1

REFERENCES

(1) SZEWCZYK, A. A. 'Combined forced and free convection laminar flow', *J. Heat Transfer, Trans. Am. Soc. mech. Engrs* 1964 **86**, 501.

(2) SPARROW, E. M. and GREGG, J. L. 'Buoyancy effects in forced convection flow and heat transfer', *J. appl. Mech., Trans. Am. Soc. mech. Engrs* 1959 **26** (1), 133.

(3) ESHGHY, S. 'Forced flow effects on free convection flow and heat transfer', *J. Heat Transfer, Trans. Am. Soc. mech. Engrs* 1964 **86**, 290.

(4) SPARROW, E. M., EICHHORN, R. and GREGG, J. L. 'Combined forced and free convection in boundary layer flow', *Physics Fluids* 1959 **2** (3), 319.

(5) ACRIVOS, A. 'Combined laminar free and forced convection heat transfer in external flows', *A.I.Ch.E. Jl* 1958 **4**, 285.

(6) SCHLICHTING, H. *Boundary layer theory* (4th edit.), 110, 121 (McGraw-Hill, New York).

(7) KREITH, F. *Principles of heat transfer*, 357 (International Textbook Co.).

(8) FALKNER, V. M. and SKAN, S. W. 'Some approximate solutions of boundary layer equations', *Phil. Mag.* 1931 **12**, 865; also *A.R.C. R. & M 1314*, 1930.

(9) ECKERT, E. R. G. 'Die Berechnung des Wärmeübergangs in der laminaren Grenzschicht Umströmter Körper', *VDI Forsch.* 1942 **416**, 1.

(10) HANS, L. S. 'Hydrodynamic entrance lengths for incompressible laminar flow in rectangular ducts', *J. appl. Mech., Trans. Am. Soc. mech. Engrs* 1960 **27** (3), 403.

C117/71

HEAT AND MASS TRANSFER BY NATURAL CONVECTION AND COMBINED NATURAL CONVECTION AND FORCED AIR FLOW THROUGH LARGE RECTANGULAR OPENINGS IN A VERTICAL PARTITION

B. H. SHAW*

Original theory for combined natural convection and forced air flow across a rectangular opening in a vertical partition has been postulated and generalized to include both heat and mass transfer. Experiments for natural convection, and combined natural convection and forced air flow, were carried out with openings 2·05 m high and from 0·10 to 0·90 m wide with air as the convecting fluid. Temperature differentials were in the order of 0–12 degC and the supply and extract volumes in the range 0·0–0·30 m³/s. Natural convection results are quoted in the range $10^8 < Gr < 10^{11}$ while the combined natural convection and forced air flow results for the Nusselt number are expressed as a function of a dimensionless group which was found to include both Reynolds and Grashof numbers.

INTRODUCTION

UNTIL 1960, studies of natural convection were primarily concerned with problems of heat transfer involving vertical and horizontal plates and bodies of varying shape. Schmidt (1)†, in a review in 1961, mentioned a type of natural convection which, until then, had received very little attention. This was the situation occurring at openings in partitions, for which Schmidt reported an optical investigation of the transient mixing of two fluids of different densities (carbon dioxide and air) separated by an opening in a vertical partition.

Apart from the transient case, the two basic aspects of natural convection through openings are those of steady conditions with vertical and horizontal partitions. Emswiler in 1926 (2) had treated the case of multiple openings in a wall and had obtained an expression for the rate of flow of air in terms of temperature difference and Bernoulli's equation for ideal flow. He did not consider the case of a single opening nor did he treat the heat and mass transfer aspects of the problem which can be generalized for all fluids.

No direct measurements had been made to substantiate or extend the theory and this may partly be explained by measurement difficulties and by the fact that opening sizes of practical importance were too large to be investigated in a laboratory. However, in the last 10 years four major sources have published theoretical and experimental work relating to this type of convection. There are a number of variables such as type of convection, area of opening, height of opening, temperature differential, and condition of opening, the combinations of which may be considered in any particular analysis. Table 117.1 compares these pertinent variables as studied by each source.

It may be seen from Table 117.1 that no previous research has been carried out with small temperature differentials. Also, there were no results for the effect, within a room, of excess pressure acting on the natural convection. The forced air flow used by Brown and Solvason was in fact a horizontal velocity parallel to the opening surface and acted as a type of air curtain. This paper does, however, consider these variables and as a result considerably widens the knowledge of convection through openings in vertical partitions.

The MS. of this paper was received at the Institution on 31st March 1971 and accepted for publication on 17th May 1971. 33

* *Building Services Research Unit, University of Glasgow, 3 Lilybank Gardens, Glasgow, W.2.*

† *References are given in Appendix 117.1.*

Notation

C	Coefficient of discharge.
C_T	Coefficient of temperature.
C_V	Coefficient of fictitious velocity.
c_p	Specific heat of fluid.
c_1, c_2	Concentration, e.g. moisture content.
D	Diffusion coefficient of mass transfer.

Table 117.1. Comparison of variables as studied in previous research

Source	Convection	Area, m^2	Height, m	Range of ΔT, degC
Brown and Solvason 1962 (3)	Natural Natural plus forced air flow	0·005 81–0·091 90	0·0762–0·3048	8–47
Graf 1964 (4)	Natural Natural plus forced air flow	—	—	Theory
Tamm 1966 (5)	Natural	—	—	Theory
Fritzsche and Lilienblum 1968 (6)	Natural	4·5	2·5	12–41·5
Shaw 1971	Natural Natural plus forced air flow	0·205–1·845	2·05	0–12

D_h Hydraulic diameter of doorway $[= 2WH/(W+H)]$.
g Acceleration due to gravity.
H Opening height.
h Heat transfer coefficient.
h_m Mass transfer coefficient.
k Thermal conductivity of fluid.
$\dot{m}$ Mass transfer rate through opening.
$\dot{q}$ Heat transfer rate through opening.
P_1, P_2 Pressures in rooms 1 and 2.
P_0 Absolute pressure at the level of the neutral zone in the opening.
P_T, P_x Pressure due to temperature differential and excess supply ventilation pressure.
Q Volumetric fluid flow rate.
Q_L, Q_x Leakage transfer volume into an area which is under positive pressure.
T_1, T_2 Temperatures in rooms 1 and 2.
t Thickness of partition.
V Velocity.
V_b Velocity defined in equation (117.19).
W Width of opening.
μ Dynamic viscosity.
ν Kinematic viscosity.
ρ Fluid density.

Dimensionless groups

Fr_Δ Densimetric Froude number $\{= V/[gH(\Delta\rho/\bar{\rho})]^{1/2}\}$.
Gr Grashof number based on density differences $[= g\,\Delta\rho H^3/\bar{\rho}\nu^2]$.
Nu Nusselt number $[= hH/k]$.
Pr Prandtl number $[= c_p u/k]$.
Re Reynolds number $[= \bar{\rho}V_b D_h/\mu]$.
Sc Schmidt number $[= \nu/D]$.
Sh Sherwood number $[= h_m H/D]$.
Sw Dimensionless group $\left[= \dfrac{Re^3}{Gr}\cdot\dfrac{H^3}{D_h{}^3} = \dfrac{\mu V_b{}^3}{\nu^2 g\,\Delta\rho}\right]$.

THEORY

Theory of the volumetric exchange of air due to natural convection through a rectangular opening in a vertical partition (3)

Consider a large sealed enclosure consisting of rooms 1 and 2 as shown in Fig. 117.1. The rooms are separated by a vertical partition with a rectangular opening of height H and width W. The temperatures in the rooms are T_1 and T_2 respectively. Since the enclosure is sealed, there is no net flow of air across the opening. The absolute pressure, P_0, at the elevation of the centre-line of the opening is everywhere equal.

In room 1, the pressure, P, at a level Z below the centre-line will be

$$P_1 = P_0+\rho_1 gZ \quad . \quad . \quad . \quad (117.1)$$

then the pressure at the same level in room 2 will be

$$P_2 = P_0+\rho_2 gZ \quad . \quad . \quad . \quad (117.2)$$

g being the acceleration due to gravity and ρ_1 and ρ_2 being the densities of air in rooms 1 and 2 respectively.

The pressure difference in these two rooms at the same level is

$$P_2-P_1 = (\rho_2-\rho_1)gZ \quad . \quad . \quad (117.3)$$

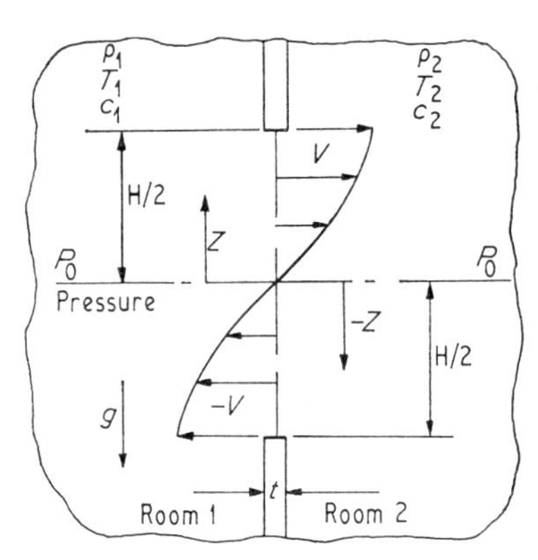

Fig. 117.1. Schematic representation of natural convection across an opening in a vertical partition

This pressure difference can be expressed as the height (h_a) of a column of air where

$$h_a = \frac{\rho_2 - \rho_1}{\bar{\rho}}; \qquad Z = \frac{\Delta\rho}{\bar{\rho}} Z$$

in which $\bar{\rho}$, the mean density, is written

$$\bar{\rho} = \frac{\rho_1 + \rho_2}{2} \quad . \quad . \quad . \quad (117.4)$$

As there is only limited information available for the relation between pressure head and velocity V for rectangular orifices at low flow rates, the flow will, in this case, be assumed to be ideal (i.e. frictionless).

For ideal flow the Bernoulli equation can be assumed, i.e.

$$V = (2gh_a)^{1/2} = \left[2g\left(\frac{\Delta\rho}{\bar{\rho}}\right)Z\right]^{1/2} \quad (117.5)$$

where V is the air velocity.

Now $Q = CAV$, where Q is the rate of volumetric discharge, C the coefficient* (unknown as yet and to be determined from tests), and A the area of the opening.

The total volumetric discharge through half of the opening can be written as

$$Q = C\int_0^{H/2} W\left[2g\left(\frac{\Delta\rho}{\bar{\rho}}\right)Z\right]^{1/2} dZ$$

On integrating this expression, the total volumetric discharge through one half of the opening will be

$$Q = C\frac{W}{3}\left[g\left(\frac{\Delta\rho}{\bar{\rho}}\right)\right]^{1/2} H^{3/2} \quad . \quad (117.6)$$

With the flow Q is now associated the heat transfer rate

$$\dot{q} = Q\bar{\rho}c_p(T_1 - T_2) \quad . \quad . \quad . \quad (117.7)$$

and the mass transfer rate, i.e. moisture content,

$$\dot{m} = Q\bar{\rho}(c_1 - c_2) \quad . \quad . \quad . \quad (117.8)$$

where c_p is the specific heat.

Introducing now the heat transfer coefficient h and the mass transfer coefficient h_m, defined as

$$h = \dot{q}/[WH(T_1 - T_2)] \quad . \quad . \quad (117.7a)$$

and

$$h_m = \dot{m}/[WH\bar{\rho}(c_1 - c_2)] \quad . \quad . \quad (117.8a)$$

equations (117.7) and (117.8) lead to the following equations in terms of dimensionless variables:

for heat transfer,

$$Nu = \frac{hH}{k} = \frac{C}{3}\left(\frac{g\,\Delta\rho H^3}{\nu^2\bar{\rho}}\right)^{1/2}\frac{c_p\mu}{k}$$

$$= \frac{C}{3}Gr^{1/2}Pr \quad . \quad . \quad . \quad . \quad . \quad (117.9)$$

for mass transfer,

$$Sh = \frac{h_m H}{D} = \frac{C}{3}\left(\frac{g\,\Delta\rho H^3}{\nu^2\bar{\rho}}\right)^{1/2}\frac{\mu}{\bar{\rho}D}$$

$$= \frac{C}{3}Gr^{1/2}Sc \quad . \quad . \quad . \quad . \quad (117.10)$$

where the symbols are as defined in the Notation.

Equations (117.9) and (117.10) cannot be exact for all conditions owing to neglect of viscosity in equation (117.5) and neglect of thermal conductivity and diffusivity in equations (117.7) and (117.8). The effect of these properties is considered in detail by Brown and Solvason (**3**). However, it is adequate to state that if air is considered to be the convecting fluid over the tested temperature differential range, the pure conduction heat transfer would be quite negligible compared with that of convection. Hence, for air in this general range, the exponents of the Grashof, Prandtl, and Schmidt numbers will not vary appreciably from those stated in equations (117.9) and (117.10).

Theory of the volumetric exchange of air due to the combined effect of natural convection and forced air flow through a rectangular opening in a vertical partition

As far as the author is aware, no theory for the above conditions has yet been written. The problem may be approached in a similar manner to that of natural convection, the only difference being that one of the rooms is under positive pressure due to air being supplied to it from an external source (Fig. 117.2). In this case the enclosures are not sealed, air being supplied to one and extracted from the other. In room 1 the pressure P, at a level Z below the centre-line, will be

$$P_1 = P_0 + \rho_1 gZ + P_x \quad . \quad . \quad (117.11)$$

where P_x is the additional pressure within the room due to the excess supply ventilation and P_0 the absolute pressure at the level of the neutral zone in the opening. The pressure at the same level in room 2 will be

$$P_2 = P_0 + \rho_2 zZ \quad . \quad . \quad (117.12)$$

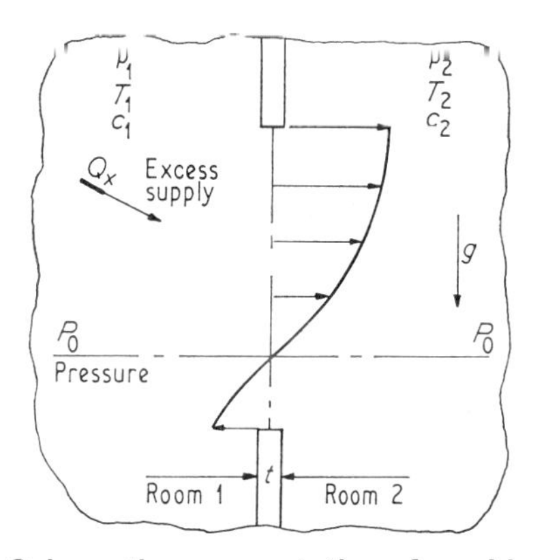

Fig. 117.2. Schematic representation of combined natural convection and forced air flow across an opening in a vertical partition

* *The coefficient C is normally referred to as the coefficient of discharge and has been taken by various sources as 0·65 for a door opening.*

The pressure difference in these two rooms at the same level is

$$P_2 - P_1 = (\rho_2 - \rho_1)gZ - P_x \quad . \quad (117.13)$$

The pressure difference and supply pressure can be expressed as the height (h_a) of a column of air where the pressure due to temperature differential is

$$h_1 = \frac{\rho_2 - \rho_1}{\bar{\rho}}; \qquad Z = \frac{\Delta\rho}{\bar{\rho}} Z$$

and the supply air pressure is

$$h_2 = \frac{P_x}{\bar{\rho}g} = \frac{V_x^2}{2g}$$

Therefore, from equation (117.13)

$$h_a = h_1 - h_2 \quad . \quad . \quad . \quad (117.14)$$

Similar limitations to that of the theory of natural convection regarding viscosity, thermal conductivity, and diffusivity must also be considered in this analysis.

The Bernoulli equation may once again be assumed, i.e.

$$V = (2gh_a)^{1/2}$$

$$= \left[2g\left(\frac{\Delta\rho}{\bar{\rho}} Z - \frac{V_x^2}{2g}\right)\right]^{1/2}$$

$$= \left[2g\left(\frac{\Delta\rho}{\bar{\rho}}\right) Z - V_x^2\right]^{1/2} \quad . \quad (117.15)$$

Now $Q_L = CAV$, where Q_L is the leakage inflow against the forced air flow. Thus,

$$Q_L = C \int_{L2}^{L_1} W \left[2g\left(\frac{\Delta\rho}{\bar{\rho}}\right) Z - V_x^2\right]^{1/2} dZ$$

where limit L_1 represents the bottom or top of the door and has the value $H/2$ since the centre-line of the door has been taken as the reference point, and L_2 is the neutral zone where supply pressure equals convective pressure occurring when $V_T^2 - V_x^2 = 0$, i.e. the pressure due to the temperature differential equals the excess supply ventilation pressure:

$$P_T - P_x = 0$$

On integrating the above expression the leakage inflow through the door will be

$$Q_L = C.W \frac{1}{2g(\Delta\rho/\bar{\rho})} \cdot \frac{2}{3} \left[2g\left(\frac{\Delta\rho}{\bar{\rho}}\right) \frac{H}{2} - V_x^2\right]^{3/2}$$

therefore,

$$Q_L = C.W \frac{1}{3} \cdot \frac{1}{g(\Delta\rho/\bar{\rho})} \left[g\left(\frac{\Delta\rho}{\bar{\rho}}\right) H - V_x^2\right]^{3/2} \quad . \quad (117.16)$$

It has already been shown that with natural convection on its own, the Nusselt number may be expressed as a function of the Grashof and Prandtl numbers, i.e. $Nu = \phi(Gr, Pr)$. This result is consistent with existing theory on natural convection. Existing theory on forced convection states that the Nusselt number may be expressed in terms of Reynolds and Prandtl numbers. As far as is known, no relationship yet exists for the combined effect of natural convection and forced air flow.

It may therefore be assumed that the Nusselt number could be expressed in terms of both natural and 'forced' convection, i.e.

$$Nu = \phi(Re, Gr, Pr)$$

The following theory proves this to be true and in the process introduces a dimensionless group which, for convenience, will be called *Sw*.

With the flow Q_L we now associate the heat transfer rate

$$\dot{q} = Q_L \bar{\rho} c_p (T_1 - T_2) \quad . \quad . \quad (117.17)$$

and the mass transfer rate

$$\dot{m} = Q_L \bar{\rho} (c_1 - c_2) \quad . \quad . \quad . \quad (117.18)$$

where c_p is the specific heat.

Introducing now the heat transfer coefficient h and the mass transfer coefficient h_m, defined as

$$h = \dot{q}/[WH(T_1 - T_2)] \quad . \quad (117.17a)$$

and

$$h_m = \dot{m}/[WH\bar{\rho}(c_1 - c_2)] \quad . \quad (117.18a)$$

equations (117.17) and (117.18) lead to the following equations in terms of dimensionless variables:

for heat transfer,

$$Nu = \frac{hH}{k} = \frac{C}{3} \cdot \frac{c_p \mu}{k} \left(\frac{\mu V_b^3}{\nu^2 g\, \Delta\rho}\right)$$

$$= \frac{C}{3} Pr.Sw \quad . \quad . \quad . \quad . \quad (117.19)$$

where V_b is the equivalent velocity within the square brackets of equation (117.16) and *Sw* is a dimensionless group.

From equation (117.19) it is found that the group *Sw* is in fact equal to the value

$$Sw = \frac{Re^3}{Gr} \cdot \frac{H^3}{D_h^3} \quad . \quad . \quad . \quad (117.20)$$

where D_h is the hydraulic diameter of the doorway. Thus the dimensionless group *Sw* is a function of the Reynolds and Grashof numbers, the height of the opening, and the hydraulic diameter of the opening. No physical meaning can be attached to this group, as can be done, for instance, with Reynolds number (ratio of inertia forces to viscous forces). However, it is nonetheless a dimensionless grouping. It can therefore be seen from this analysis that with combined natural convection and forced air flow, the Nusselt number can be represented by

$$Nu = \frac{C}{3} Pr \frac{Re^3}{Gr} \cdot \frac{H^3}{D_h^3} \quad . \quad . \quad (117.21)$$

For mass transfer a similar analysis may be carried out leading to the following expression:

$$Sh = \frac{h_m H}{D} = \frac{C}{3} Sc.Sw = \frac{C}{3} Sc \frac{Re^3}{Gr} \cdot \frac{H^3}{D_h^3} \quad (117.22)$$

EXPERIMENTAL

Test area

The test area was situated at the Experimental Ward Unit at Hairmyres Hospital, East Kilbride, the tests being carried out in the isolation rooms of the intensive care area (Fig. 117.3). The rooms opened into a common air lock/vestibule. Radiators were positioned in each of the three rooms to supply additional heating to that of the supply air, and a sheet of expanded polystyrene was placed over the window in room 'A' to reduce any heat loss through the window.

Instrumentation

The mechanical supply and extract volumes to each room were measured and balanced with averaging pressure flowmeters in accordance with Ma's method of balancing (7). Air temperatures in the room and doorways were measured using copper–constantan thermocouples. Hot wire anemometers were used to measure the air velocities in the doorway. Air flow direction through the doorways was initially determined using cotton wool swabs soaked in titanium tetrachloride, but it was found that the smoke propagated rusting and cigarette smoke was subsequently used to determine the air direction.

Scope of tests and procedure

Tests were conducted for single rectangular openings of the following nominal dimensions: 0·90, 0·50, and 0·10 m wide, all areas being 2·05 m high. These different door areas were set up by blanking off the door openings with wooden boards. Supply and extract volumes to the rooms were varied from 0·05 to 0·30 m^3/s in steps of 0·05 m^3/s. Balanced ventilation systems (natural convection) had equal supply and extract volumes while the positive ventilation systems (combined natural convection and forced air flow) had only supply air. Air temperature differences ranged from 0 to 12 degC. Owing to the massiveness of the test apparatus periods of up to 3 hours were required to reach equilibrium conditions, especially when large temperature differentials were being set up.

When the air temperatures within each area had stabilized, a grid consisting of a vertical Meccano strip was suspended from the top of each opening. These grids had 10 anemometers and 10 thermocouples fixed at equal intervals down their length and were suspended in such a manner that the air velocities and temperatures at any vertical section could be measured. Air flow direction at each point on the grid was then determined and the anemometer heads adjusted accordingly to face the oncoming air flow. If the direction was not definite, i.e. in the neutral zone, the anemometer head was placed sideways. The anemometer and thermocouple readings were then recorded for that particular vertical section, five sets of readings being taken and averaged. When recording had finished the grids were moved to their next position and the procedure repeated.

In order to obtain a useful picture of the air movement through the opening and the air temperatures at the openings, three vertical grid position readings were obtained for the 0·90 m opening, two positions for the 0·50 m, and one position for the 0·10 m. Once this procedure had been completed for a specific opening area, the wooden boards were placed in the doorways to reduce the area to the required dimensions. The whole test procedure was then repeated for the new opening areas.

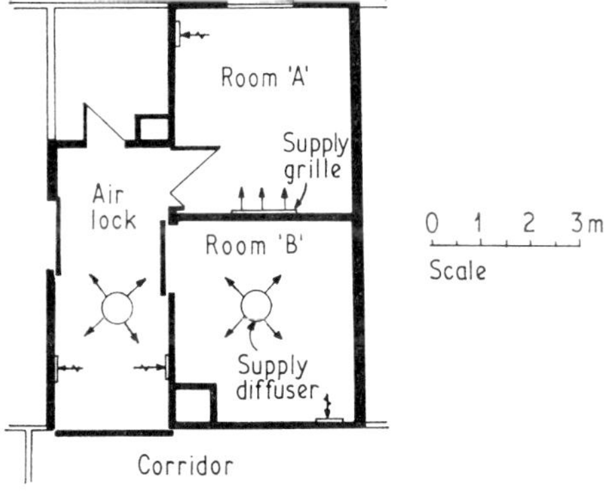

Fig. 117.3. Plan of test area

TEST RESULTS

The velocity readings that had been recorded during the tests were analysed with a trend surface analysis program which fitted the best curve (linear, quadratic, and cubic) to the results, thus calculating the volume of air flowing in and out through the opening. The program also printed out isovel diagrams of the air movement in the doorway.

Natural convection

The temperature differential which was used in the analysis of the results was that of the temperature difference between top and bottom of the opening. This was thought to be the most appropriate differential with respect to the theory. The air temperature used in determining the dimensionless groups was that of the average of the top and bottom temperatures at the opening, thus giving a mean heat transfer coefficient.

The coefficient of discharge values for natural convection were obtained by dividing the actual convective transfer volume (from the test results) by the basic theoretical volume, i.e. equation (117.6). The coefficient values were found to be primarily a function of temperature differential, the door area not being significant (Fig. 117.4). It was therefore decided to refer to the coefficient as the coefficient of temperature. An interesting feature of Fig. 117.4 is that from about 4 degC downwards the value of the coefficient increases asymptotically with the coefficient axis. The reason for this trend may be explained as follows. The convective transfer volumes at zero temperature differential (by extrapolation) for each door area are listed in Table 117.2. By dividing these values by half the door area a mean velocity may be obtained. This results in a mean velocity of 0·1362 m/s (27 ft/min) for any door area. As the free air velocity, or turbulence, within the ventilated room is generally quoted as being in

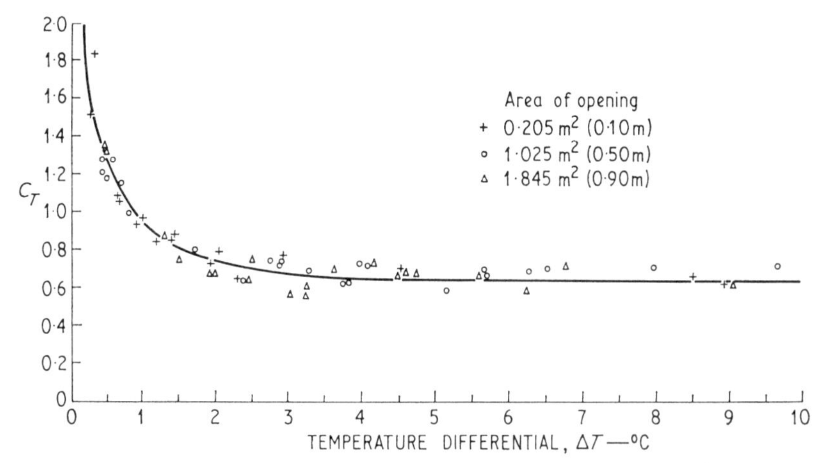

Fig. 117.4. Coefficient of temperature (C_T)

Table 117.2. Transfer volumes at zero temperature differential

Door area, m²	Transfer volume, m³/s	Mean velocity, m/s
1·845	0·1250	0·1360
1·025	0·0700	0·1362
0·205	0·0140	0·1362

the range 0·1016–0·1524 m/s (20–30 ft/min), this strengthens the validity of the experimental results and explains why the coefficient does not remain constant at 0·65.

Above 10 degC temperature differential the coefficient rises again, very slowly this time, reaching a value of unity at about 50 degC differential and continuing to rise. This is not, in fact, shown in Fig. 117.4. Although the reason for this trend is not at present apparent, it compares favourably with limited results of Fritzsche and Lillienblum (**6**) working in the region of 20–30 degC differential.

Experimental results for heat transfer are given in Fig. 117.5, where the Nusselt number divided by the Prandtl number is ordinate and the Grashof number is abscissa. For comparison and verification of the theory, results of Brown and Solvason (**3**) are also shown. With an opening 2·05 m high, the upper theoretical curve, and a temperature differential of 10 degC, the Grashof number equals $1{\cdot}3\times10^{10}$. As can be seen, this is the point where the theoretical curve breaks away from the broken line (coefficient of 0·65). For further reference, at differential 40 degC, $Gr = 6{\cdot}76\times10^{10}$. Comparing these values with

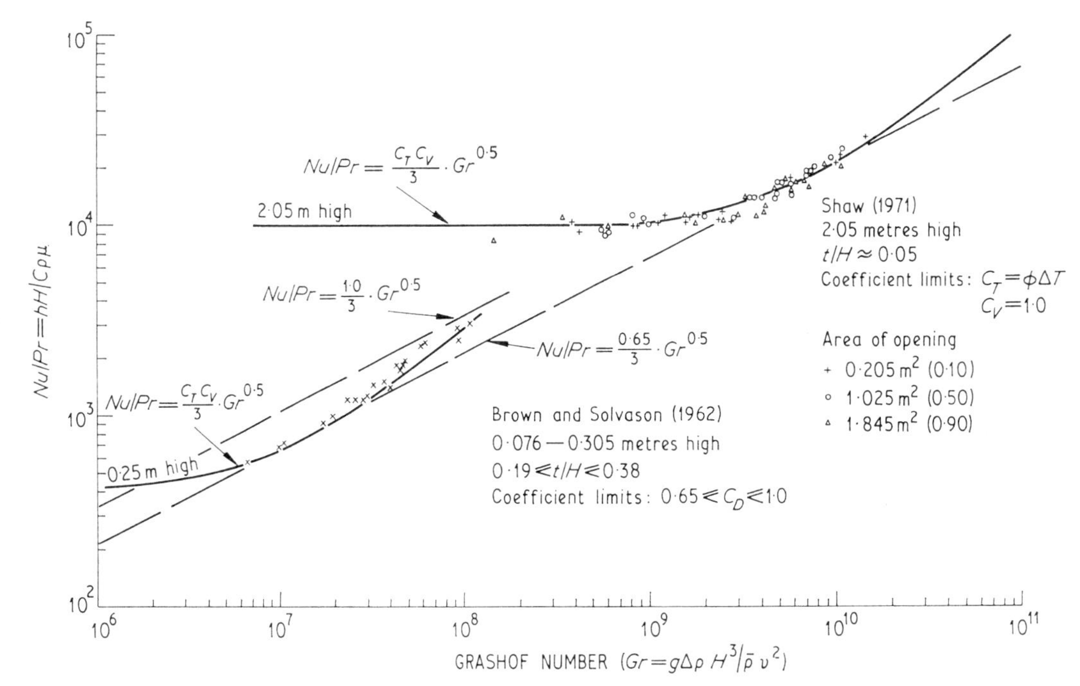

Fig. 117.5. Natural convection across rectangular openings in a vertical partition ($10^6 < Gr < 10^{11}$)

the results of Brown and Solvason it may be seen that a theoretical curve for an opening 0·25 m in height correlates favourably with their results. It is now possible to verify the validity of the theoretical curves for higher differentials in the range 10–40 degC, as this was the region in which Brown and Solvason were working. With this height of opening and a differential of 10 degC, $Gr = 2{\cdot}35 \times 10^7$ while at 40 degC, $Gr = 1{\cdot}23 \times 10^8$. This is in fact the breakaway region from the lower broken line and is similar to the upper curve. The two broken lines bounding the lower results are the limits of coefficient values stated by Brown and Solvason (0·65–1·0).

The type of flow for these tests is considered to be turbulent. Also, since the Prandtl number for air in the range of temperatures used in the tests was constant at 0·71, it was not possible to investigate its influence as a separate variable.

The theory for natural convection stated above is consistent with the approach of previous workers, as can be seen from the list below. There are, however, two points which differ between sources, these being reference density and the use of a coefficient.

Brown and Solvason (1962):

$$Q = C_D \frac{W}{3} \cdot H^{3/2} (g)^{1/2} (\Delta\rho/\bar{\rho})^{1/2}$$

Graf (1964):

$$Q = \frac{W}{3} \cdot H^{3/2} (g)^{1/2} (\Delta\rho/\bar{\rho})^{1/2}$$

Tamm (1966):

$$Q = \frac{W}{3} H^{3/2} (g)^{1/2} (\Delta\rho/\rho_c)^{1/2}$$

Fritzsche and Lilienblum (1968):

$$Q = C_T \frac{W}{3} H^{3/2} (g)^{1/2} (\Delta\rho/\rho_c)^{1/2}$$

Shaw (1971):

$$Q = C_T . C_V \frac{W}{3} H^{3/2} (g)^{1/2} (\Delta\rho/\bar{\rho})^{1/2}$$

The existence of an excess pressurization coefficient C_V was not at first apparent as the coefficient had a value of unity for natural convection, i.e. a balanced ventilation scheme with no excess pressure.

Combined natural convection and forced air flow

On analysis of the positive tests in conjunction with the balanced tests, which may be regarded as positive tests with no excess supply pressure, it became evident that another coefficient did in fact exist. By dividing the actual inflow volume by the theoretical inflow volume—equation (117.16) without a coefficient of discharge—an overall coefficient was obtained, this being a function of a fictitious velocity over the area of the opening due to excess supply (Q_x/A) and the temperature differential, i.e.

$$C = \phi(Q_x/A, \Delta T)$$

where Q_x is equal to the supply volume minus the extract volume. The discharge coefficient, C, was found to be a product of the temperature coefficient, C_T, as obtained from the balanced tests, and a fictitious velocity coefficient, C_V, i.e.

$$C = C_T \times C_V$$

By dividing the left-hand side of this equation by the temperature coefficient it was possible to find the values of the fictitious velocity coefficients. The value of this coefficient started at unity for a system with no excess volume, decreasing as the amount of excess volume increased (Fig. 117.6).

Experimental results for heat transfer are shown in Fig. 117.7, the ordinate once again being Nu/Pr while the abscissa is the dimensionless group Sw. Broken lines in this graph represent temperature differential while full lines represent fictitious air velocity over the opening due to excess supply pressure (Q_x/A). As expected, the excess pressure reduces the heat transfer rate across the opening. Once again the type of flow is considered to be in the turbulent regime.

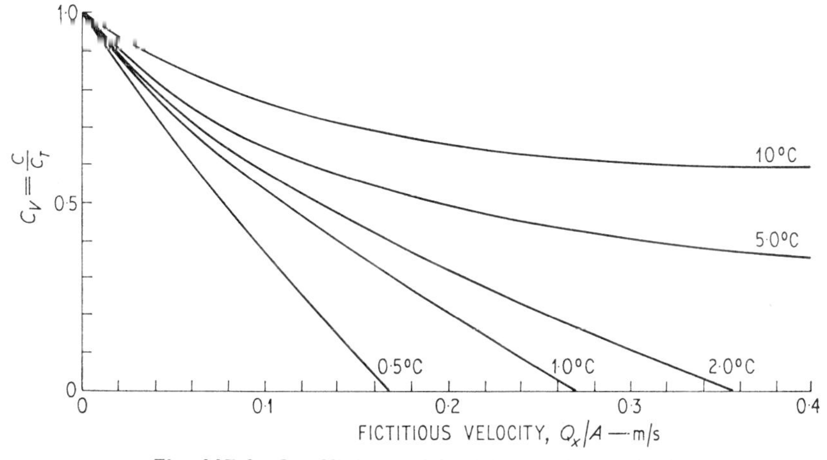

Fig. 117.6. Coefficient of fictitious velocity (C_V)

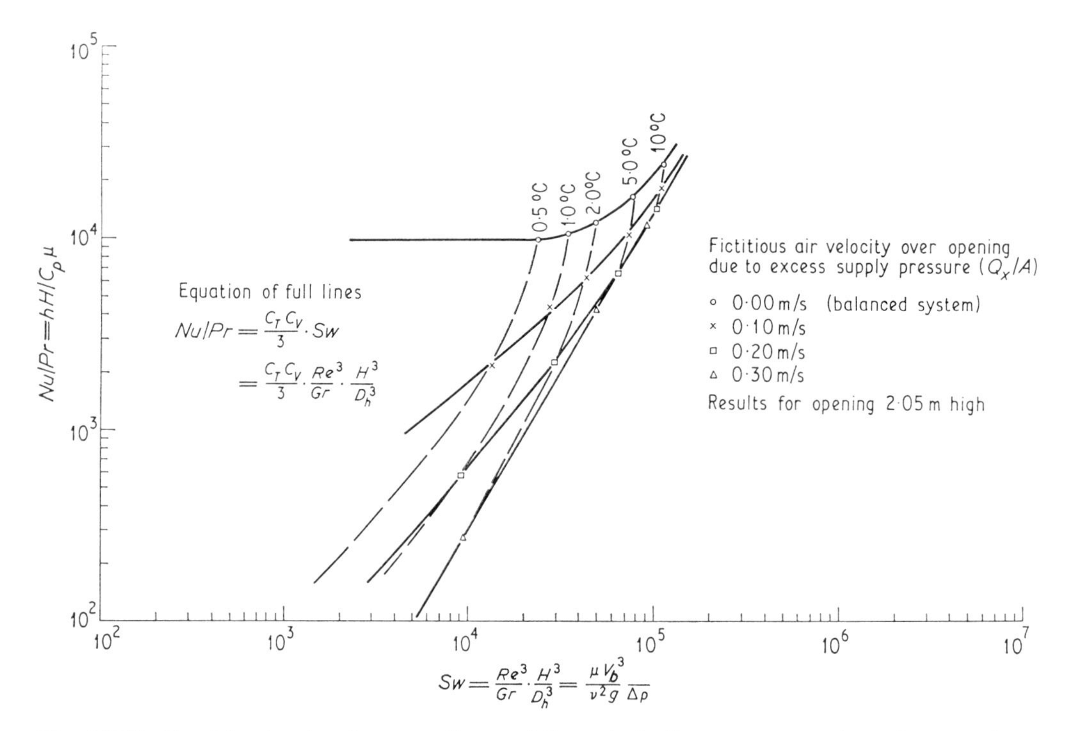

Fig. 117.7. Combined natural convection and forced air flow across rectangular openings in a vertical partition ($10^3 < Sw < 10^4$)

DISCUSSION AND CONCLUSIONS

The test results for air in natural convection, and combined natural convection and forced air flow, across rectangular openings in vertical partitions are in good agreement with theory.

It must be remembered that the theory is not exact for all conditions owing to neglect of viscosity, thermal conductivity, and diffusivity in certain equations. However, as stated previously, if air is considered to be the convecting fluid over the tested temperature differential ranges then the theory may be used with confidence. The type of flow is considered in both cases to be turbulent.

Since the characteristic dimensionless numbers for natural convection and forced convection may be taken as the Grashof and Reynolds numbers, respectively, it was assumed that the characteristic dimensionless group for combined natural convection and forced air flow would be a function of both Grashof and Reynolds numbers. This was indeed found to be the case, the relationship being in the ratio, $Sw = (Re^3/Gr)(H^3/D_h^3)$, where Sw is a dimensionless group and the other significant terms are the height and hydraulic diameter of the opening.

An interesting analogy to the subject of this paper is the densimetric exchange flow of water in rectangular channels. A lock gate or other such division may separate bodies of still water of the same surface level but which differ slightly in density. This density difference may be due to either temperature or salinity differential. Barr (**8**), in a paper on this subject, expressed this mechanism in terms of a densimetric Froude–Reynolds number, which is a criterion involving differential gravitational and viscous forces, i.e.

$$Fr_{\Delta} . Re = \left(\frac{g\,\Delta\rho H^3}{\bar{\rho}\nu^2}\right)^{1/2} \quad . \quad (117.23)$$

This is equivalent to natural convection with air as the convecting fluid, and it is interesting to note that the right-hand side of equation (117.23) is in fact the Grashof number $(Gr^{1/2})$ of equation (117.9), i.e. the characteristic dimensionless group. Barr does not introduce a forced flow on the natural exchange, but it may be noted that the dimensionless group for this mechanism may also be expressed in terms of the densimetric Froude number and Reynolds number, i.e.

$$Sw = \frac{Re^3}{Gr} \cdot \frac{H^3}{D_h^3} = \left(\frac{\mu V_b^3}{\nu^2 g\,\Delta\rho}\right) = Re . Fr_{\Delta}^2 \frac{H}{D_h} \quad (117.24)$$

There exist several fields of study to which the results of this paper may be applied. First there is the problem in public buildings, shops, supermarkets, and restaurants of convective air currents causing unpleasant draughts and loss of heat at doorways where where is normally a heavy concentration of pedestrian traffic. The general movement of air within buildings is also of importance whether it be naturally ventilated or air conditioned, and under special conditions, such as a fire within the building, it is essential to know the movement of smoke up vertical shafts—especially in tall buildings. The problem concerned with cold storage rooms, with temperature differentials in the range 30–40 degC, is that of heat and mass transfer through the access doorway resulting in greater running costs. To counteract these losses, air screens and mechanically operated doors are used, yet they still form a large part of the heat balance of many cold storage depots whose actual amount should be a matter of precise knowledge both for the planning engineer and the manager of cold stores. With the theory and results of this paper it is now possible to determine accurately the volumetric exchange of air, hence heat and mass transfer, for all these situations.

The importance of the airborne route of infection in critical areas within hospitals has been shown by numerous workers for several decades. In particular, various papers on the subject of convective transfer through doorways of bacteria such as *Staph. aureus*, an important group of bacteria which causes wound infection, have been published in the last 10 years. It is now possible to predict the volumetric exchange through doorways under certain conditions and, hence, the isolation efficiency of the system. It is also possible to determine the amount of excess ventilation required to completely isolate the critical area from the rest of the hospital.

ACKNOWLEDGEMENTS

The author would like to thank Messrs W. Carson and W. Whyte and members of the Building Services Research Unit for advice and guidance, and Sister M. F. D. Muir and the staff of the Hairmyres Experimental Ward for their co-operation. This work was sponsored by the Department of Health and Social Security.

APPENDIX 117.1

REFERENCES

(1) SCHMIDT, E. 'Heat transfer by natural convection', *Int. Heat Transfer Conf.*, University of Colorado, 1961.

(2) EMSWILER, J. E. 'The neutral zone in ventilation', *Trans. Am. Soc. Heat. Vent. Engrs* 1926 **32**, 59.

(3) BROWN, W. G. and SOLVASON, K. R. 'Natural convection through rectangular openings in partitions—vertical partitions', *Int. J. Heat Mass Transfer* 1962 **5**, 859.

(4) GRAF, A. 'Consideration of the air exchange between two rooms', *Schweiz. Bl. Heiz. Lüft.* 1964 **31** (1), 22.

(5) TAMM, W. 'Cold losses through openings in cold rooms', *Kaltetechnik-Klimatisierung* 1966 **18** (42), 142.

(6) FRITZSCHE, C. and LILIENBLUM, W. 'New measurements for the determination of cold losses at the doors of cold rooms', *Kaltetechnik-Klimatisierung* 1968 **20** (9), 279.

(7) MA, W. Y. L. 'The averaging pressure tubes flowmeter for the measurement of the rate of airflow in ventilation ducts and for balancing of air-flow circuits in ventilating systems', *J. Instn Heat. Vent. Engrs* 1967 (No. 34, February), 327.

(8) BARR, D. I. H. 'Densimetric exchange flow in rectangular channels', *Houille blanche* 1963 (No. 7), 739.

C118/71

COMBINED FREE AND FORCED CONVECTION HEAT TRANSFER IN A VERTICAL PIPE

J. E. BYRNE* E. EJIOGU†

Measurements of heat transfer coefficient in upward flow in a vertical pipe show a large decrease from the values expected for a horizontal pipe with turbulent flow when $Gr/Re^{2\cdot8} = -10^{-4}$. This criterion has been established theoretically by considering the effect of buoyancy forces on turbulence production in the pipe and confirmed by measurement of velocity and turbulence intensity profiles. However, if the ratio $|Gr/Re^{1\cdot8}| \leqslant 0\cdot05$ then the pipe flow will be unaffected by the buoyancy forces.

INTRODUCTION

THE EXPERIMENT TO BE DESCRIBED is one in which buoyancy forces are allowed to modify a turbulent flow in a vertical pipe. Such forces do, of course, always exist in a heated fluid provided it has a finite coefficient of expansion and is acted upon by a gravitational force field. However, it is not usual for them to be significant unless the Reynolds number of the forced flow is small.

Buoyancy effects can, nevertheless, be quite important in a number of engineering applications in which the Reynolds number is high ($\sim 10^5$). It is merely necessary that the Grashof number should be large enough—a situation which occurs readily with fluids having a large coefficient of expansion (e.g. a fluid close to its critical point) or in systems of large dimensions. A number of interesting effects on forced convection heat transfer have been observed with supercritical pressure fluids, and have been attributed primarily to buoyancy forces rather than to variations in transport properties (**1**)‡. It is therefore of interest to see whether similar effects can be observed with normal fluids in which the variation of transport properties is small but where buoyancy effects are made important by choosing a pipe of large diameter. With air at atmospheric pressure in a 600-mm diameter pipe, Grashof numbers can be achieved which are similar to those known to produce significant modifications to heat transfer in supercritical pressure carbon dioxide in a 20-mm diameter pipe. These considerations formed a basis for the choice of conditions in the experiment to be described.

Few investigations into the effect of buoyancy forces on forced convection heat transfer appear to have been made under conditions of turbulent flow at moderate and high Reynolds numbers (10^4–10^5). This is perhaps not surprising when one considers the scale of apparatus required for the case of a gas at atmospheric pressure. Smaller scale apparatus will suffice in the case of liquids, but even then, pipes of approximately 100 mm in diameter are necessary. Brown and Gauvin (**2**), reporting measurements of temperature fluctuations in water in a vertical pipe, state that the interaction of an upward forced flow in a heated pipe with a free convective flow can lead to lower heat transfer coefficients than expected. Similar effects are also observed with supercritical pressure fluids, and in this case the reduction in heat transfer coefficient can be very large (**3**). The sense of the effect thus appears to be at variance with the common assumption that heat transfer is improved by mixed convection provided that the velocities generated by forced and free flows are in the same direction.

Many theoretical analyses of non-turbulent mixed convection flows have been made, the more recent involving numerical solution of the equations of motion and energy. In general, they have confirmed the expectation that heat transfer is improved when the forced and free flows are in the same direction (e.g. laminar flow of a heated fluid flowing upwards in a vertical pipe). The difficulty in extending such analyses to turbulent flow lies in the fact that buoyancy forces, in radically modifying the shape of the shear stress distribution across the pipe, can affect the turbulence production and the turbulent diffusivity. In the present experiments the flow was always turbulent at the point where heating commenced, and even though the turbulence was sometimes reduced by heating, the flow probably remained turbulent throughout most of the pipe.

The MS. of this paper was received at the Institution on 21st April 1971 and accepted for publication on 25th May 1971. 23

* *University of Manchester, M13 9PL.*

† *Foster Wheeler John Brown Boilers Ltd, P.O. Box 160, Greater London House, Hampstead Rd, London, N.W.1.*

‡ *References are given in Appendix 118.1.*

Notation

- a Pipe radius.
- c_p Specific heat at constant pressure before start of heating.
- f Friction factor $[= \tau_0/\frac{1}{2}\rho_m u_m^2]$.
- Gr Grashof number $\{= [8g\rho_m(\rho_m-\rho_0)a^3]/\mu^2\}$.
- g Gravitational acceleration.
- h Heat transfer coefficient based on wall to bulk temperature difference.
- k Thermal conductivity before start of heating.
- Nu Nusselt number $[= 2ha/k]$.
- Pr Prandtl number $[= c_p\mu/k]$.
- Re Reynolds number $[= 2\rho_m u_m a/\mu]$.
- T Temperature.
- u Velocity in the x direction.
- u^* Dimensionless velocity $[= u/u_m]$.
- v Velocity in the y direction.
- v^* Dimensionless velocity $[= v/u_m]$.
- x Distance along the pipe.
- x^* Dimensionless distance $[= x/a]$.
- y Distance in radial direction from wall to centre of pipe.
- y^* Dimensionless distance $[= y/a]$.
- y^+ Dimensionless distance $\{= [y(\rho_m\tau_0)^{1/2}]/\mu\}$.
- μ Viscosity before start of heating.
- ρ Fluid density.
- τ Shear stress.
- τ^* Dimensionless shear stress $[= \tau/\rho_m u_m^2]$.

Subscripts

- a Centre-line values.
- c Values for pure forced convection only.
- m Mean values.
- p Values at the position of maximum production.
- 0 Wall values.

FORMULATION OF THE PROBLEM

Consider flow in a vertical pipe in which all properties but density remain constant. Even in the case of density, variations can be ignored except in so far as they give rise to density difference terms (i.e. 'buoyancy' terms) in the equation of motion. Under these conditions the momentum equation in the flow direction becomes

$$\rho u\frac{\partial u}{\partial x}+\rho v\frac{\partial u}{\partial y}=\frac{1}{a-y}\cdot\frac{\partial}{\partial y}\{\tau(a-y)\}+\frac{2\tau_0}{a}-g(\rho_m-\rho) \quad . \quad . \quad . \quad (118.1)$$

or, in dimensionless form,

$$u^*\frac{\partial u^*}{\partial x^*}+v^*\frac{\partial u^*}{\partial y^*}=\frac{1}{1-y^*}\cdot\frac{\partial}{\partial y^*}\{\tau^*(1-y^*)\}+2\tau_0^* -\frac{1}{2}\cdot\frac{Gr}{Re^2}\frac{(\rho_m-\rho)}{(\rho_m-\rho_0)} \quad (118.2)$$

In the present experiment the initial condition is that of a fully developed turbulent flow in which the buoyancy term is zero. Equation (118.2) then reduces to

$$\frac{1}{1-y^*}\cdot\frac{\partial}{\partial y^*}\{\tau^*(1-y^*)\}+2\tau_0^*=0 \quad . \quad (118.3)$$

which is satisfied by a linear shear stress distribution across the pipe, varying from τ_0^* at the wall to 0 at the centre-line. Thus the importance of the buoyancy term may now be assessed by comparing its magnitude with τ_0^*. Thus, buoyancy effects will be significant when

$$\left|\frac{Ge}{Re^2}\cdot\frac{\rho_m-\rho}{\rho_m-\rho_0}\right| \sim 2\tau_0^*$$

Noting that the maximum value of $(\rho_m-\rho)/(\rho_m-\rho_0) = 1$, the criterion becomes

$$\left|\frac{Gr}{Re^2}\right| \sim 2\tau_0^* = f$$

where f is the friction factor, which for smooth pipe turbulent flow is $f = 0{\cdot}046Re^{-0{\cdot}2}$. Thus, buoyancy will affect the flow when

$$\left|\frac{Gr}{Re^{1{\cdot}8}}\right| \sim 0{\cdot}05 \quad . \quad . \quad . \quad (118.4)$$

CONDITIONS FOR MINIMUM HEAT TRANSFER

In laminar flow the effect of the buoyancy force is to increase the heat transfer by increasing the velocity near the wall. However, in turbulent flow the change in shear stress caused by the buoyancy force (Fig. 118.1) can affect the turbulence structure and, since turbulence accounts for the relatively higher rates of heat transfer, the possibility exists of a deterioration in heat transfer coefficient due to a 'laminarization' of the near wall layer. If a laminar layer in such a case occurs in the region of an originally turbulent pipe flow where maximum turbulence

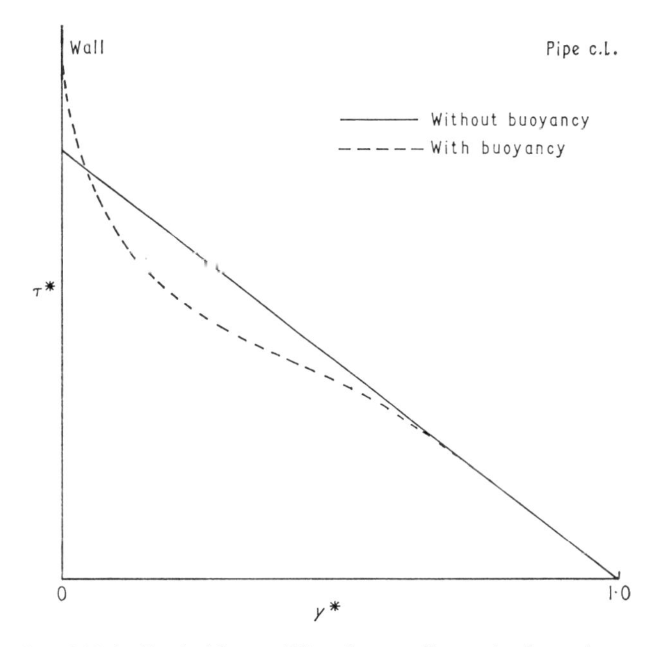

Fig. 118.1. Probable modification to dimensionless shear stress distribution by buoyancy forces in upward flow

production would normally take place, then between the wall and this region the laminar relationship between shear stress and velocity gradient in non-dimensional form is

$$\tau^* = \frac{2}{Re}\cdot\frac{\partial u^*}{\partial y^*}$$

and approximating the shear stress distribution by a linear equation which satisfies the local boundary conditions only, i.e.

$$\tau^* = \tau_0^* - \left(\tau_0^* - \frac{1}{2}\cdot\frac{Gr}{Re^2}\right) y^* \quad . \quad (118.5)$$

this shear stress will become zero at a point y_p^*, given by the equation

$$\tau_0^* = -\frac{1}{2}\cdot\frac{Gr}{Re^2}\,\frac{y_p^*}{1-y_p^*} \quad . \quad . \quad (118.6)$$

The value of the velocity at this point y_p^* is

$$u_p^* = -\frac{Gr}{8Re}\,y_p^{*2}\,\frac{(1+2y_p^*)}{(1-y_p^*)} \quad . \quad (118.7)$$

Now if y_p^* coincides with the value of y^*, where production of turbulence was a maximum when no heating was applied to the pipe, then production will cease in the region $0 \leqslant y^* \leqslant y_p^*$. Since this production depends on the velocity gradient present, it will be small in the remainder of the pipe provided $u_p^* \simeq 1$, given that the velocity gradient around $y^* = y_p^*$ is already small or even zero if the flow is laminar near the wall.

y_p^* can be considered small compared with unity for the flows under consideration here, and defining

$$y^+ = y^*\,\frac{Re}{2}\left(\frac{f}{2}\right)^{1/2}$$

where f is the friction factor for normal pipe flow, the expression

$$\frac{Gr}{Re^3} = -\frac{u_p^* f}{y^{+\,2}_{p}} \quad . \quad . \quad . \quad (118.8)$$

is obtained using equations (118.5), (118.6) and (118.7).

Now the centre of maximum production in a pipe flow is normally around $y^+ \simeq 20$, and using the previous expression for friction factor, the criterion for minimum heat transfer for turbulent flow in a pipe with a large buoyancy force is

$$\frac{Rr}{Re^{2\cdot 8}} \simeq -10^{-4} \quad . \quad . \quad . \quad (118.9)$$

DESCRIPTION OF APPARATUS

This consisted essentially of a closed loop of 0·613-m diameter pipe arranged vertically in a stair well of the Simon Engineering Laboratories (Fig. 118.2). It provided a vertical pipe some 35 m in height, of which the top 3·7 m was heated. A variable speed axial flow fan and a cooler were situated at the bottom of the loop. Reynolds numbers in excess of 10^5 could be achieved.

The heated length was split into five sections, four of which were used as preheaters, the uppermost forming the test section. As shown in Fig. 118.2, the test section was split into 30 rings 22·3 mm high, thermally insulated

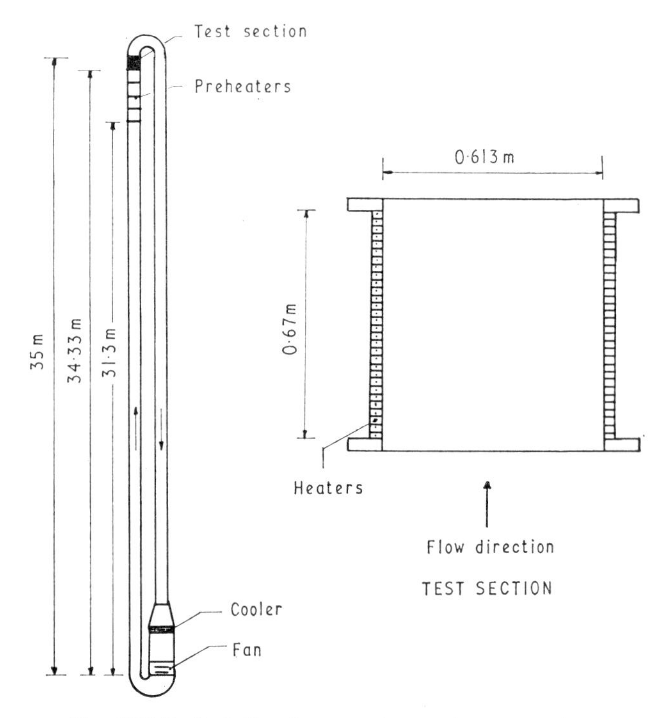

Fig. 118.2. Details of apparatus and test section

from each other by thin layers of insulation, and fitted with individual heaters and thermocouples, thus enabling the thermal boundary conditions to be controlled. In the present experiments the heated length was maintained at a uniform temperature, and the distribution of heat flux in the test section could then be determined from the heat inputs to the individual rings.

The flow in the pipe was measured by means of a hot-wire traverse immediately below the start of the heated length. Temperatures of the pipe wall were measured using 130 thermocouples distributed along the heated length. In the test section three thermocouples were distributed around each ring, i.e. 90 thermocouples in all. It was found possible to achieve uniformity of temperature over the test section to within ±0·5 degC, and the variation over the preheaters was maintained within ±3 degC. The heated section of the pipe was well lagged, and a separate measurement of heat losses from the exterior of the test section was made.

Velocity, temperature, and turbulence traverses were carried out using a 0·025-mm diameter constant temperature hot-wire anemometer at the outlet of the test section. Different overheat ratios could be used to separate out the mean velocity and temperature signals from the hot wire. It cannot be claimed that these measurements are sufficiently accurate to form a basis of a detailed analysis of the flow. The experimental problems were severe because of the low velocities and, in the case of turbulence measurements, the interaction between temperature and velocity fluctuations. However, it is felt that the traverses are useful as a qualitative indication of the changes produced by the buoyancy forces.

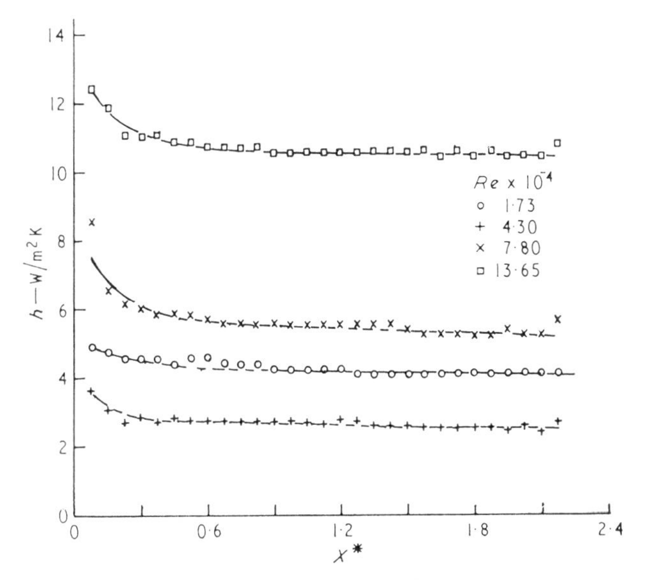

Fig. 118.3. Distribution of heat transfer coefficient along the test section for various Reynolds numbers and a fixed Grashof number of $1{\cdot}34 \times 10^9$

EXPERIMENTAL RESULTS

Heat transfer coefficients

Local heat transfer coefficients for the test section were computed from the measured power inputs, wall temperatures, and local bulk temperatures (calculated by means of a heat balance from the start of the heated section to the point in question). In view of the relatively short heated length, the change in fluid temperature was usually small compared with the difference between the pipe wall and the fluid.

The variation in heat transfer coefficients along the test section is shown in Fig. 118.3 for a range of Reynolds numbers and one wall temperature. The variation is small, and it is thought that the apparent rise at inlet (corresponding to an x^* calculated from the start of heating of 9·1) may be due to locally greater heat losses from the flange connecting the test section to the preheater. In view of the small variation with x^*, the remaining results are presented as average heat transfer coefficients for the test section.

Fig. 118.4 shows the Nusselt number (based on the average heat transfer coefficient for the text section) as a function of Reynolds number for a range of Grashof numbers. Also marked in the figures is the computed Nusselt number for the case of negligible buoyancy effects. This has been estimated using the expression

$$Nu = 0{\cdot}021 Re^{0{\cdot}8} . Pr^{0{\cdot}6}$$

modified to allow for thermal entrance effects (4). Limiting values of Nu for zero forced flow could not be determined because under these conditions a stagnant zone of hot air was established in the upper part of the pipe loop and temperatures continued to rise since the only mechanism of heat removal was by losses from the outer walls of the pipe.

Velocity, temperature, and turbulence traverses

Figs 118.5, 118.6, and 118.7 show velocity, temperature, and turbulence measurements across the pipe at the outlet of the test section for one value of Grashof number. Once

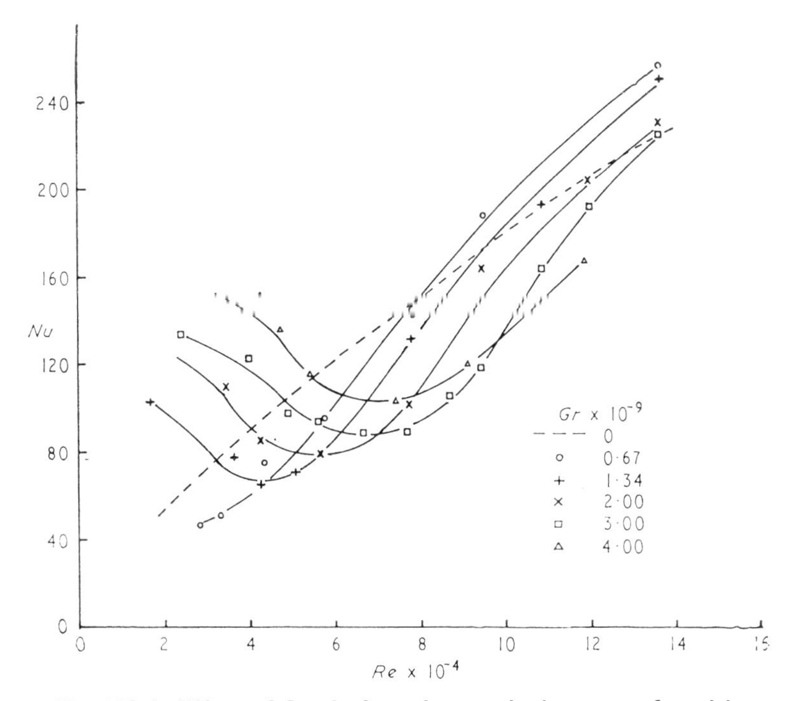

Fig. 118.4. Effect of Grashof number on the heat transfer with forced convection

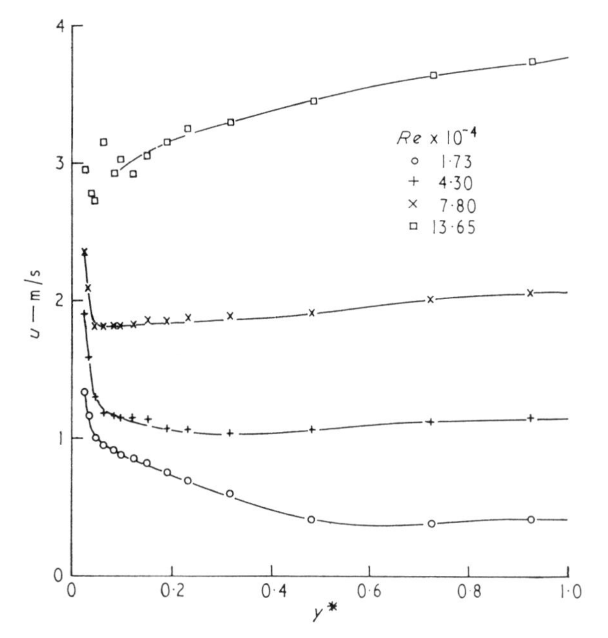

Fig. 118.5. Velocity distribution across the pipe for a Grashof number of $1{\cdot}34 \times 10^9$ at various Reynolds numbers

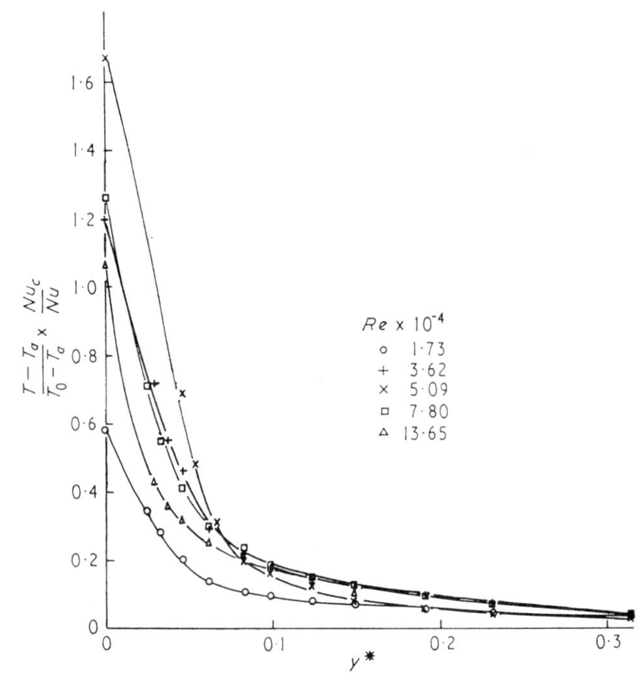

Fig. 118.6. Dimensionless temperature profiles near the pipe wall for a Grashof number of $1{\cdot}34 \times 10^9$ and various Reynolds numbers

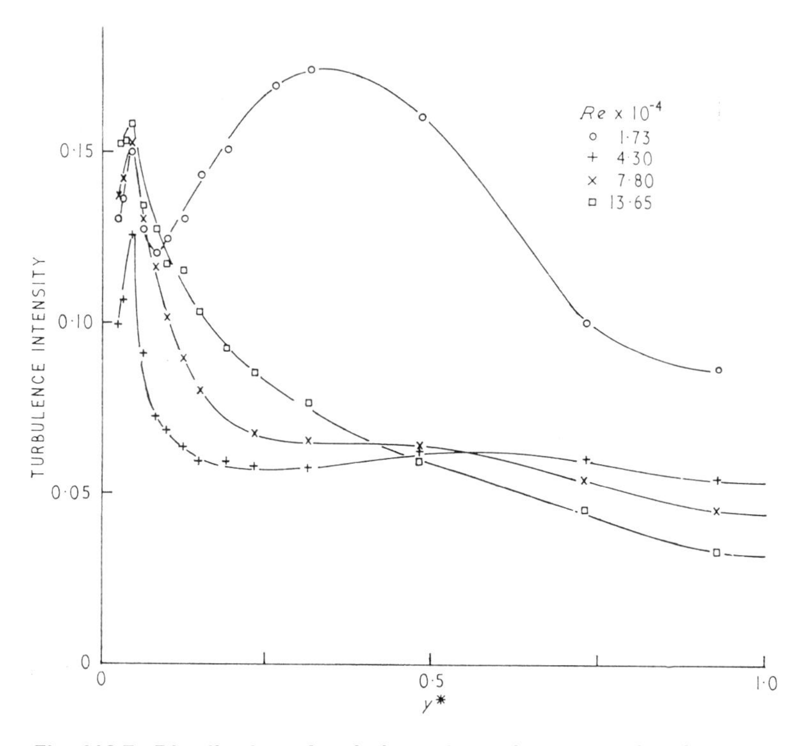

Fig. 118.7. Distribution of turbulence intensity across the pipe at a Grashof number of $1{\cdot}34 \times 10^9$ and various Reynolds numbers

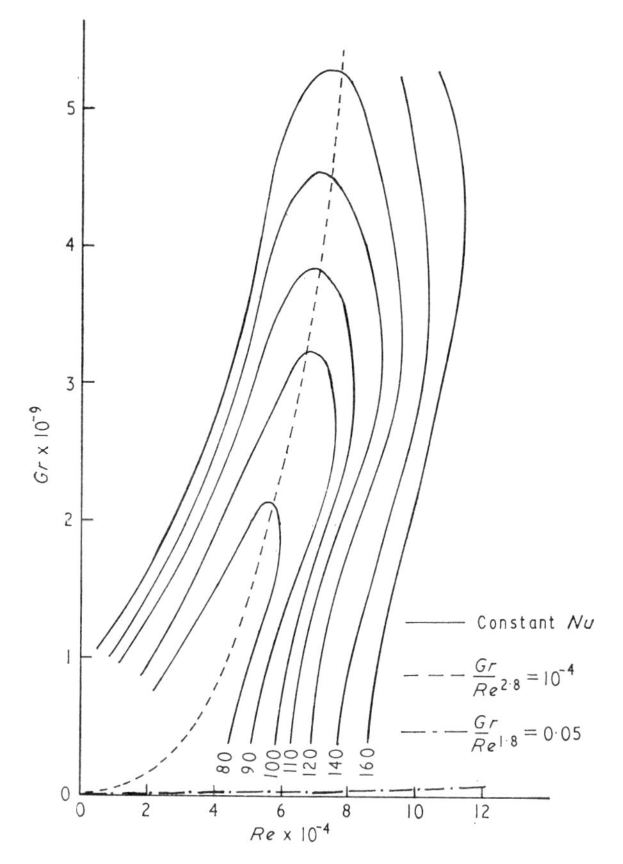

Fig. 118.8. Contours of constant Nusselt number on a grid of Grashof and Reynolds numbers

again it is necessary to mention that in view of the experimental difficulties the measurements of time mean velocity and temperature are unlikely to have an absolute accuracy of more than ±10 per cent. However, they are intended primarily to indicate the changes that occur when buoyancy forces become important. No attempt has been made to separate the turbulent temperature and velocity fluctuations, but since the hot-wire temperature was about 210°C (apart from the region close to the wall), the velocity fluctuation will probably dominate. Because the temperature and velocity fluctuations are fairly highly correlated, the combined signal will still be representative of the change in turbulence caused by buoyancy forces.

DISCUSSION OF RESULTS

Fig. 118.4 shows that the effect of buoyancy forces on a turbulent heated upward flow is generally to reduce the heat transfer coefficient below that for the case of no buoyancy forces. At very low flows, or at high Grashof numbers, free convection dominates the situation and the Nusselt number may exceed that for forced convection. The minimum value of Nusselt number occurs at a value of Reynolds number which increases as the Grashof number increases.

The results shown in Fig. 118.4 are re-plotted in Fig. 118.8 in the form of lines of constant Nusselt numbers on a grid of Grashof and Reynolds numbers. It will be seen that the region of minimum heat transfer corresponds closely to the theoretical criterion $|Gr/Re^{2\cdot 8}| = 10^{-4}$. The criterion for the buoyancy forces having some effect [equation (118.4)] is also shown in this figure, and it appears to be a very conservative estimate. Indeed, there is a small (~5 per cent) effect up to $|Gr/Re^{1\cdot 8}| = 0{\cdot}5$, i.e. 10 times the original estimate.

It is not easy to assess the effects of buoyancy using the normal dimensionless temperature distributions, since there is a Reynolds number effect present in addition to the changes caused by buoyancy. The results have therefore been plotted in Fig. 118.6 in the form of the dimensionless temperature multiplied by Nu_c/Nu, where Nu is the experimental Nusselt number and Nu_c is the Nusselt number for pure forced convection at the same Reynolds number. The ordinate may therefore be interpreted as the local temperature difference $(T - T_a)$ made dimensionless by dividing by the wall to centre-line temperature difference for pure forced convection at the same heat flux, i.e. $(T_0 - T_a)(Nu/Nu_c)$. It will be seen that, apart from the case of $Re = 1{\cdot}73 \times 10^4$ where free convection dominates, there is an increase in temperature drop in the region $0 \leqslant y^* \leqslant 0{\cdot}1$ which almost vanishes at $Re = 1{\cdot}365 \times 10^5$ where forced convection dominates.

The velocity distributions (Fig. 118.5) do not, unfortunately, give sufficient detail close to the wall to locate the peak velocity in this region. They do, however, show that at the higher Reynolds numbers the flow outside the region $y^* > 0{\cdot}1$ does not appear to be greatly affected by buoyancy, and has a typical turbulent forced flow shape. For the Grashof number corresponding to the results in Fig. 118.5, the minimum heat transfer coefficient would correspond to a Reynolds number of about $4{\cdot}5 \times 10^4$. Close to the wall the turbulence production, which is proportional to the turbulent shear stress and the gradient of the time mean velocity, might also be expected to pass through a minimum at this Reynolds number. (In a flow without buoyancy the region of maximum turbulence production occurs at $y^* \simeq 0{\cdot}02$ for this Reynolds number.)

The distribution of turbulent intensity (Fig. 118.7) confirms to some extent the above hypothesis. Taking the values at the higher Reynolds numbers as being characteristic of a flow that is little affected by buoyancy, the turbulent intensity in the wall region for the Reynolds number corresponding to the minimum value of heat transfer coefficient is reduced.

CONCLUSIONS

The experimental results show that the interaction of a forced flow and a free convective flow under turbulent conditions can result in a reduced heat transfer coefficient when compared with a purely forced flow. The minimum heat transfer coefficient is obtained when the group $Gr/Re^{2\cdot 8}$ is approximately equal to -10^{-4}, and buoyancy has a negligible effect on the forced flow when $|Gr/Re^{1\cdot 8}| < 0{\cdot}05$.

It must be borne in mind that the experimental results presented here apply to a relatively short heated length, although the flow was almost fully developed at entry to this length.

A probable mechanism for the reduction in heat transfer is the creation of a laminar layer near the wall due to the modification to the shear stress by the buoyancy forces and the consequent reduction in the turbulence production in the whole flow.

ACKNOWLEDGEMENT

The authors gratefully acknowledge the advice and assistance of Professor W. B. Hall.

APPENDIX 118.1

REFERENCES

(1) HALL, W. B. and JACKSON, J. D. 'Laminarization of a turbulent pipe flow by buoyancy forces', *A.S.M.E.–A.I.Ch.E. Heat Transfer Conf.*, Minneapolis, 1969.

(2) BROWN, C. K. and GAUVIN, W. H. 'Temperature profiles and fluctuations in combined free and forced convection flows', *Chem. Engng Sci.* 1966 **21**, 961.

(3) HALL, W. B., JACKSON, J. D. and WATSON, A. 'A review of forced convection heat transfer to fluids at supercritical pressures', *Symp. Heat Transfer and Fluid Dynamics of Near Critical Fluids, Proc. Instn mech. Engrs* 1967–68 **182** (Pt 3I), 10.

(4) KUTATELADZE, S. S. and BORISHANSKII, V. M. *A concise encyclopedia of heat transfer* 1966 (Pergamon, Oxford).

C119/71

LAMINAR COMBINED NATURAL AND FORCED CONVECTION IN A RECTANGULAR FIELD

A. P. HATTON* N. H. WOOLLEY*

The numerical solution of the steady two-dimensional equations of motion and energy was carried out for the case of developing Couette flow with a heated patch on the stationary wall. The three cases of upflow, downflow, and horizontal flow were calculated for a range of Reynolds number from 0 to 2000 and of Grashof number from 0 to 5×10^6. Contour plots of streamlines and isotherms were obtained, and these show that under certain conditions the flow separates near the heated patch with both downflow and horizontal flow and an associated eddy is formed. Mean Nusselt numbers for downflow show a minimum value which can be much lower than for pure forced convection.

INTRODUCTION

BUOYANCY FORCES may significantly affect the forced convection heat transfer in a number of practical situations. For example, in high-pressure boilers and in reactor cooling channels, with large temperature differences present, these effects may give rise to departures from the unheated flow pattern. This behaviour of the flow, in turn, causes significant errors if forced convection predictions are used. The experimental examination of these effects is difficult and costly and it is worthwhile to call on available computer techniques in the hope that this will provide useful theoretical evidence.

In practice the flow is usually turbulent but prediction methods in this field are in an early stage of development. For duct and boundary layer flows turbulent predictions are achieving considerable success (1)† but much development is yet required for the more complex situations in which flow reversals are present. Simple turbulent flows have been successfully calculated by assuming the turbulent fluid to behave as a non-newtonian laminar fluid in which the 'viscosity' may be a function of mean velocity gradient, distance from the wall, conditions upstream, and other parameters.

Before such ideas can be extended into complex flows, however, it is essential to develop computing methods on the simpler constant viscosity case.

The object of this work was to investigate the computational techniques available for laminar flows with a view to their extension into the turbulent situation. However, it is hoped that the results themselves may be of some practical interest since the effects of natural convection are more significant in low-speed flows.

The MS. of this paper was received at the Institution on 4th May 1971 and accepted for publication on 21st May 1971. 33

* *University of Manchester Institute of Science and Technology, P.O. Box 88, Manchester M60 1QD.*

† *References are given in Appendix 119.1.*

Notation

A	Band matrix of coefficients.
C	Specific heat.
C_f	Friction factor $[= \tau_w/\tfrac{1}{2}\rho_0 U^2]$.
Gr	Grashof number $\left[= \dfrac{T_1 - T_0}{T_0} \cdot \dfrac{gl^3}{\nu^2}\right]$.
g	Acceleration due to gravity.
h	Local heat transfer coefficient.
h_a	Mean heat transfer coefficient.
K_ϕ	Right-hand side vector of equation for ϕ.
k	Thermal conductivity.
l	Length of heated patch.
Nu_a	Mean Nusselt number $[=h_a l/k]$.
P	Pressure.
Pr	Prandtl number.
Re	Reynolds number $[= Ul/\nu]$.
St	Local Stanton number $[=h/\rho_0 CU]$.
T	Local absolute temperature.
T_0	Ambient absolute temperature.
T_1	Absolute temperature of heated patch.
U	Characteristic inlet velocity.
u	Component of velocity in the x direction.
v	Component of velocity in the y direction.
x, y	Co-ordinate system directions.
α	Thermal diffusivity.
θ	Dimensionless temperature.
ν	Kinematic viscosity.
ρ	Local density.
ρ_0	Ambient density.

τ_w Wall shear stress.
ϕ General dependent variable.
ψ Stream function.
ω Vorticity.

Subscripts

exit At the exit of system.
m Maximum value.
n Normalized variable.
s At the start of heated patch.

A prime denotes a dimensionless variable.

THE PROBLEM INVESTIGATED

Because of the complex nature of the equations it was necessary to choose a simple geometry to permit solutions to be obtained using reasonable computation times. Accordingly we chose to examine a developing Couette flow with a co-ordinate system allowing varying step lengths in both directions. The stationary wall contained a heated length at uniform temperature and the flow field considered was taken large enough to assume the flow direction to be parallel to the wall at inlet and with negligible streamline curvature at outlet (Fig. 119.1). With a constant inlet velocity this case corresponds approximately to a boundary layer flow. It is simple to modify the program to deal with flow between stationary walls or to deal with other geometries, some of which are illustrated in Fig. 119.2. In all these cases uniform wall temperature or varying wall temperature can be included and, of course, pure forced or pure natural convection form two special simple cases. Any inclination to the horizontal of these passages is a further modification which is easily introduced.

In certain simple cases, for example with forced upward flow in a parallel channel, it may be possible to solve the problem by more straightforward methods than those used in this work. In general, however, the equations form an elliptic system of coupled non-linear partial differential equations which require an iterative method of solution. Such solutions are at present under intensive development for many fluid flow situations and some success has been achieved both for laminar and turbulent flows (2). To date, however, buoyancy forces have been excluded from these computations.

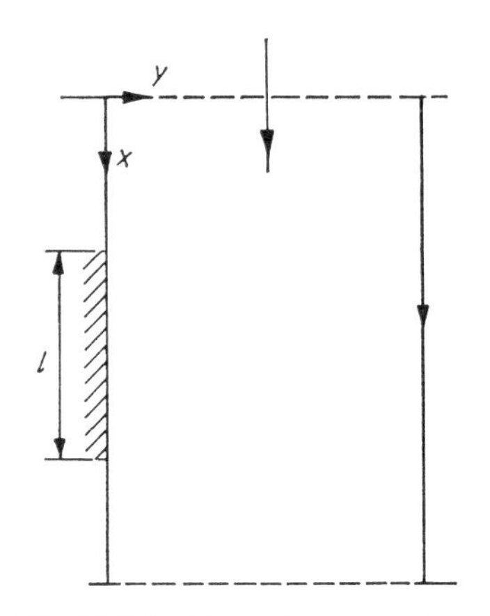

Fig. 119.1. Co-ordinate system

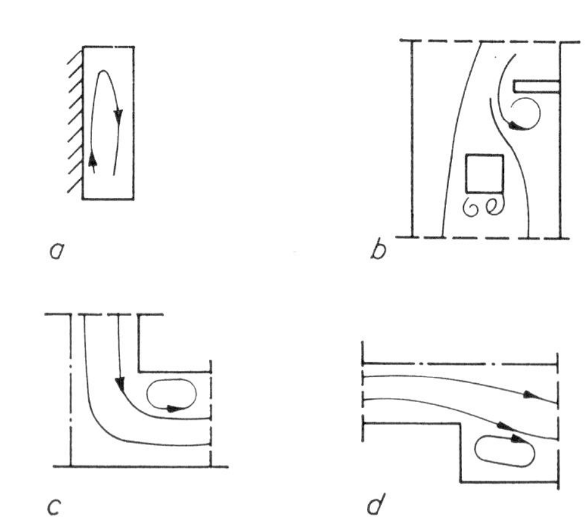

Fig. 119.2. Some possible boundary conditions

PREVIOUS WORK

All the previous theoretical studies have considered a forced convection situation in which the velocity profile was modified by the influence of the buoyancy forces, but in no case was the study carried through to include flow reversal. Rao and Morris (3) considered a fully developed situation in a parallel wall channel and showed that the heat transfer with downward flow was reduced as the Rayleigh number increased. Acrivos (4) considered boundary layer solutions using profile assumptions and established limits for the parameter Gr/Re^2 outside which only one of the two mechanisms would predominate. Gunness and Gebhart (5) describe a wedge flow solution and gave particular attention to the separation wedge case which was affected by buoyancy forces.

Experimental studies were carried out by Brown and Gauvin (6) on upward and downward flow in pipes. They showed that with a downward turbulent flow the heat transfer increased with Rayleigh number—a contrary effect to that observed in laminar flow. They also pointed out the rapid transition to turbulence caused by the instability of the situation. Hatton *et al.* (7) experimented with flows over horizontal cylinders and showed that with downward flow at the fixed Rayleigh number the heat transfer first decreased then increased as the Reynolds number increased. This, of course, was due to the flow change from upward by natural convection to downward by forced convection.

THE BASIC EQUATIONS

In order to achieve a solution it is necessary to make certain simplifying assumptions:

(*a*) The motion is steady and two dimensional.

(*b*) The fluid is incompressible in the sense that the motion does not cause density variation and also that the density variations do not affect the motion except through the buoyancy force.

(*c*) All other physical properties are assumed constant.

With these restrictions the equations become:

Momentum

in the x direction

$$u\frac{\partial u}{\partial x}+v\frac{\partial u}{\partial y}=-\frac{1}{\rho_0}\cdot\frac{\partial P}{\partial x}+\frac{g(\rho-\rho_0)}{\rho_0}+\nu\left[\frac{\partial^2 u}{\partial x^2}+\frac{\partial^2 u}{\partial y^2}\right] \quad . \quad . \quad . \quad (119.1)$$

in the y direction

$$u\frac{\partial v}{\partial x}+v\frac{\partial v}{\partial y}=-\frac{1}{\rho_0}\cdot\frac{\partial P}{\partial y}+\nu\left[\frac{\partial^2 v}{\partial x^2}+\frac{\partial^2 v}{\partial y^2}\right] \quad (119.2)$$

Continuity

$$\frac{\partial u}{\partial x}+\frac{\partial v}{\partial y}=0 \quad . \quad . \quad . \quad (119.3)$$

Energy (temperature)

$$u\frac{\partial T}{\partial x}+v\frac{\partial T}{\partial y}=\alpha\left[\frac{\partial^2 T}{\partial x^2}+\frac{\partial^2 T}{\partial y^2}\right] \quad . \quad (119.4)$$

The stream function, ψ, and the vorticity, ω, are introduced and defined by $u=\partial\psi/\partial y$ (hence, by continuity, $v=-\partial\psi/\partial x$) and $\omega=\partial v/\partial x-\partial u/\partial y$. In order to eliminate the pressure gradient term we follow the usual procedure of differentiating equation (119.1) with respect to y and equation (119.2) with respect to x and subtracting equation (119.2) from (119.1) which yields

$$\frac{\partial\psi}{\partial x}\cdot\frac{\partial\omega}{\partial y}-\frac{\partial\psi}{\partial y}\cdot\frac{\partial\omega}{\partial x}=\frac{\partial}{\partial y}\left[\frac{g(\rho-\rho_0)}{\rho_0}\right]-\nu\left[\frac{\partial^2\omega}{\partial x^2}+\frac{\partial^2\omega}{\partial y^2}\right]$$

The equations are then made dimensionless by the following substitutions. Let U be a characteristic velocity, e.g. the inlet velocity, and l be a characteristic length, e.g. the length of the heated patch. Put $y'=y/l$, $x'=x/l$, $\psi'=\psi/Ul$.

$$\psi_n=\text{normalized value of } \psi'=\frac{\psi'}{\psi'_m}$$

where ψ'_m is the maximum value of ψ' at the inlet

$$\omega'=\omega\left(\frac{l}{U}\right)$$

$$\theta=\frac{T-T_0}{T_1-T_0}$$

$$\frac{\rho-\rho_0}{\rho_0}=-\left(\frac{T-T_0}{T_0}\right)=-\theta\left(\frac{T_1-T_0}{T_0}\right)$$

Omitting the primes for clarity, the equations then become

$$\psi_m\frac{\partial\psi_n}{\partial y}\cdot\frac{\partial\omega}{\partial x}-\psi_m\frac{\partial\psi_n}{\partial x}\cdot\frac{\partial\omega}{\partial y}=\frac{Gr}{Re^2}\frac{\partial\theta}{\partial y}+\frac{1}{Re}\left(\frac{\partial^2\omega}{\partial x^2}+\frac{\partial^2\omega}{\partial y^2}\right) \quad . \quad . \quad . \quad (119.5)$$

$$\omega=-\psi_m\left(\frac{\partial^2\psi_n}{\partial x^2}+\frac{\partial^2\psi_n}{\partial y^2}\right) \quad . \quad . \quad (119.6)$$

$$\psi_m\frac{\partial\psi_n}{\partial y}\cdot\frac{\partial\theta}{\partial x}-\psi_m\frac{\partial\psi_n}{\partial x}\cdot\frac{\partial\theta}{\partial y}=\frac{1}{Re.Pr}\left(\frac{\partial^2\theta}{\partial x^2}+\frac{\partial^2\theta}{\partial y^2}\right) \quad (119.7)$$

These form a system of three coupled non-linear PDEs with dependent variables θ, ψ, and ω.

BOUNDARY CONDITIONS

It is fairly simple to specify the boundary conditions at inlet and on each side of the domain. At the exit, however, the situation is not straightforward. We chose to assume that the exit plane was sufficiently distant for the curvature of the streamlines and the decay rate of the temperature profile to be small and negligible. The choice of a uniform inlet velocity and a particular streamline parallel to the wall is intended to be an approximation to boundary layer flow.

The boundary conditions for this case are:

at $y=0$: $\quad \psi_n=\dfrac{\partial\psi_n}{\partial y}=0$

$\theta=1$ on the heated patch and

$\theta=0$ over the remainder

at $y=y_m$: $\quad \psi_n=1, \quad \dfrac{\partial\psi_n}{\partial y}=\dfrac{1}{y_m}, \quad \theta=0$

at $x=0$: $\quad \psi_n=\dfrac{y}{y_m}, \quad \theta=0$

at $x=x_{\text{exit}}$: $\quad \dfrac{\partial^2\psi_n}{\partial x^2}=0, \quad \dfrac{\partial\theta}{\partial x}=0$

The buoyancy force was changed in sign for upward and downward flow, while for horizontal flow the buoyancy term in equation (119.5) was altered to $(Gr/Re^2)(\partial\theta/\partial x)$.

METHOD OF SOLUTION

The first numerical method used was to convert the equations into algebraic equations using three-point truncated Taylor series and central differences throughout. These equations were then solved by successive point iteration (relaxation). It was not possible to obtain convergence with this scheme at Reynolds numbers greater than 60. It was realized that increased Reynolds numbers resulted in larger values of the off-diagonal elements of the equivalent matrix which reduced the 'diagonal dominance' necessary for convergence in the relaxation method.

A more direct method of solution was therefore attempted which, it was hoped, would overcome this difficulty. Linear equations can be solved by direct matrix inversion without the necessity for diagonal dominance, and it was hoped that this method, when applied to the non-linear case, would enable convergence to be achieved at higher Reynolds numbers.

The equations were all written in the form

$$\nabla^2\phi=K_\phi$$

where ϕ refers to vorticity, stream function, or temperature. Using five-point central difference expressions to reduce truncation errors the equations were transformed into sets of algebraic equations of the form

$$A\phi=K_\phi$$

where A is a banded matrix of coefficients common to the

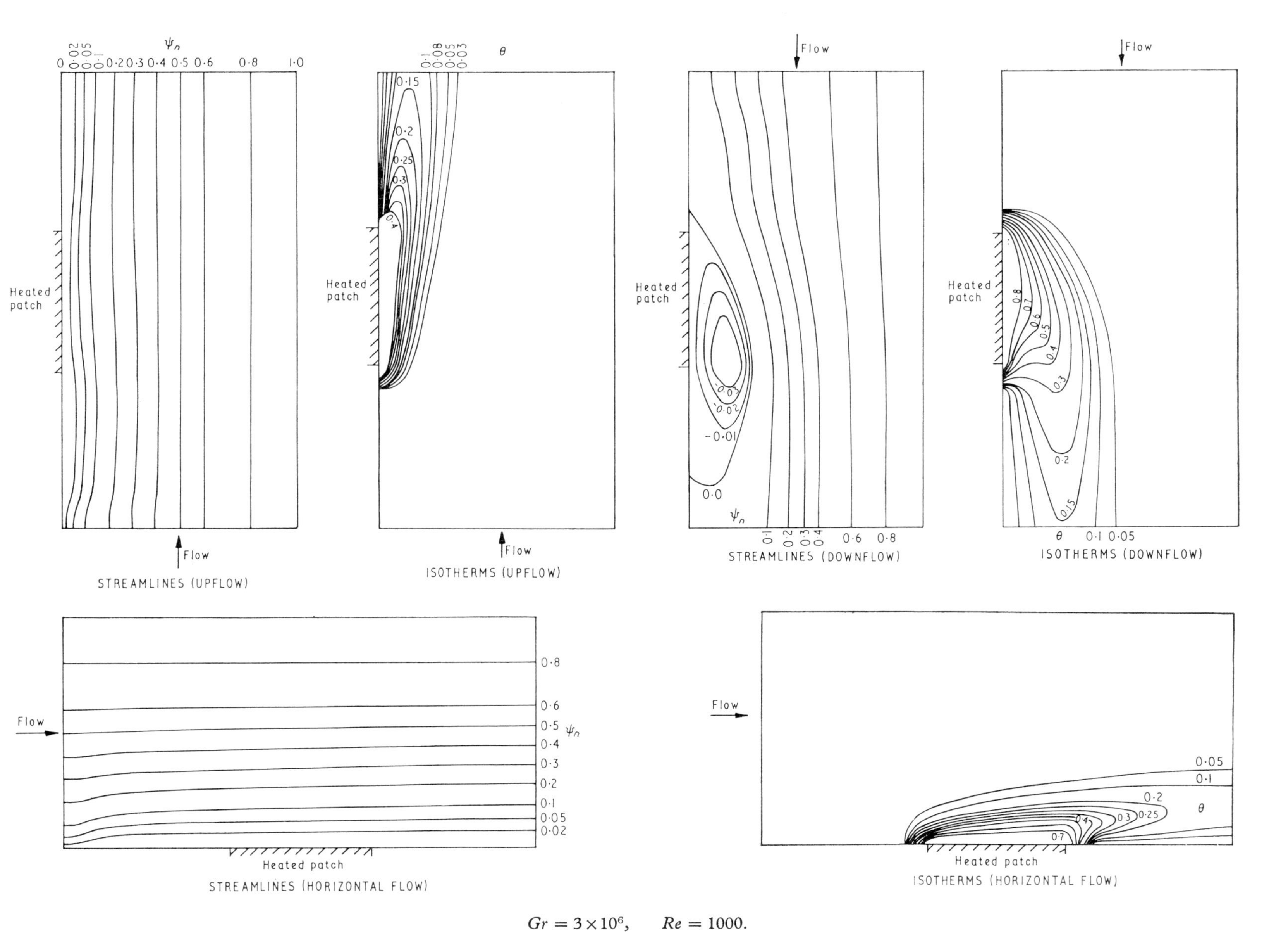

$Gr = 3\times10^6$, $Re = 1000$.

Fig. 119.3. Streamlines and isotherms for the three flow directions

three sets. Initial guesses of all the variables enable the three K vectors to be obtained and the solution of the ϕ variables carried out using the Thomas algorithm (a form of Gauss elimination). This enabled a new set of K vectors to be determined and this iteration was continued until the solution converged.

This method, however, proved to be no more convergent and no faster than relaxation. Variations of the method, such as using other terms in the equations to form the matrix, were more time consuming and no more successful. For a simple flow case, Couette flow with no buoyancy, divergence was obtained at higher Reynolds numbers even when the correct solution was supplied as the initial guess.

It was then realized that diagonal dominance is a major factor in convergence whether matrix or relaxation methods are used. The work of Spalding *et al.* (**2**) has emphasized the importance of the 'upwind difference' technique in securing stability. In fact this method has the property of ensuring diagonal dominance of the equivalent matrix.

The essential feature of this method is that finite difference expressions for the first derivatives of ω and θ are formed from the values at the point under consideration and the neighbouring upstream point. The local stream direction is determined from the ψ values around the point in question. It is not easy to include this modification in a matrix technique whereas it is simple to include in the relaxation method. Accordingly this modification was made and the calculation was then found to converge rapidly, and apparently without any upper limit on Reynolds number.

The numerical solution was carried out using a rectangular mesh chosen, in these calculations, as 21 by 21. The program allowed for the use of relaxation factors for each variable, but only the value of unity has been used in this work.

The calculation was continued until the maximum change over one iteration of the three variables was 0·5 per cent of the largest value. Local Stanton numbers and friction factors were obtained by three-point differentiation formulae and mean values by numerical integration.

The values of the variables at every point could be

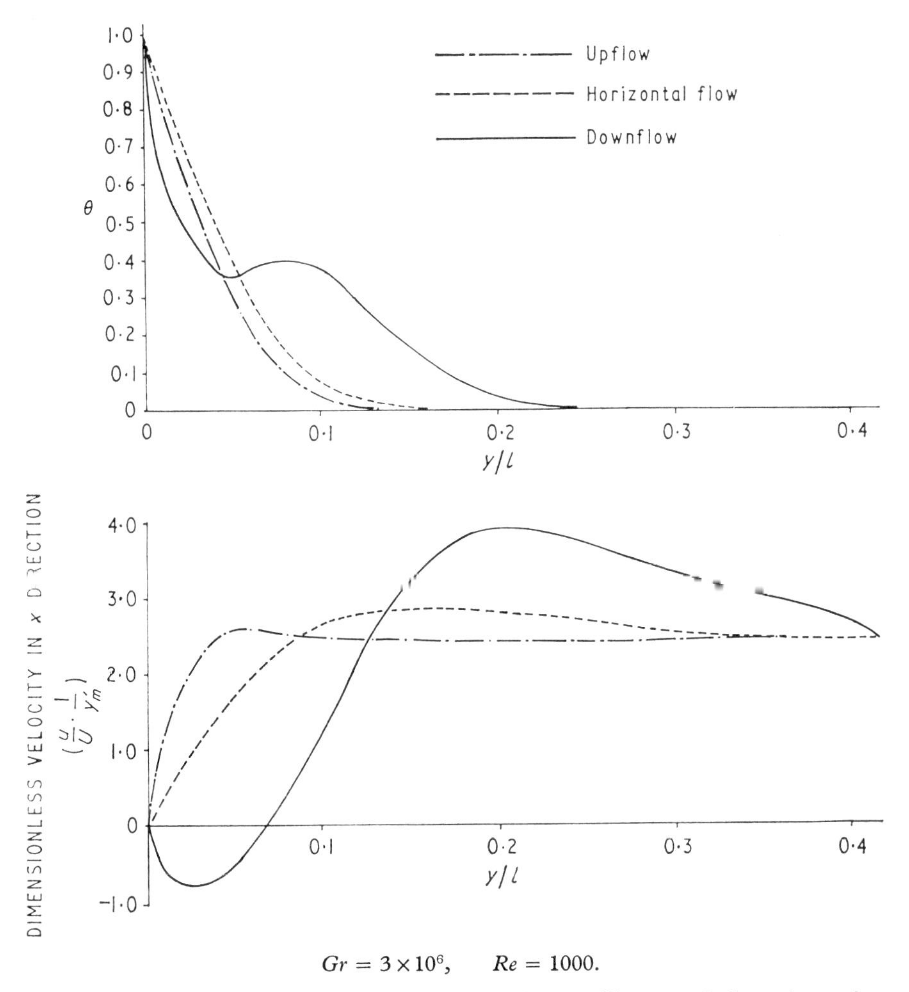

$Gr = 3\times10^6$, $Re = 1000$.

Fig. 119.4. Dimensionless temperature and velocity profiles at exit from heated patch

printed or produced on paper tape for use with an associated contour plotting program which used linear interpolation between values at the mesh points.

The number of iterations required to reach a solution increased with larger Grashof numbers. Typically for a Grashof number of 3×10^6 the number of iterations was 45, which corresponded to 90 seconds on the Manchester University Atlas computer.

DISCUSSION OF RESULTS

The calculation was carried out for a range of Reynolds number from 6 to 2000 and Grashof number from zero to 5×10^6. Only the case of Prandtl number unity was examined in detail. Particular attention was paid to determining the situations in which flow separation occurred. One such case ($Re = 1000$, $Gr = 3\times10^6$) is shown in Fig. 119.3 in which isotherms and streamlines obtained from the contour plotting program are shown. Note that these diagrams are not to the same scale in each direction. The cross-stream scale has been enlarged by a factor of 4. The boundary values chosen on the stationary wall were zero on the unheated mesh points and unity on the heated points. The contour plotting program interpolated values on the boundary over the finite step at each end of the heated portion. In the horizontal and upward flows the heating has little influence on the streamline shape although the acceleration caused by buoyancy is apparent in the upward case. The effect also causes the reduction in thickness of the thermal boundary layer, which can be seen from the isotherms. For the particular conditions shown on Fig. 119.3 the downflow case shows a recirculating region in which the buoyancy forces predominate. There are, of course, situations of lower Grashof number where there is no recirculation, but results were obtained which showed that separation may occur in horizontal flows.

Fig. 119.4 shows calculated profiles of temperature and velocity at the downstream edge of the heated patch where 'downstream' refers to the main flow. The profiles do not show any unusual features apart from the distortions caused by the recirculating flow case. As an approximation to boundary layer flow the chosen position of the straight streamline would appear to be adequate for horizontal and upward flow, but could be improved for downward flow as indicated by the non-asymptotic approach to the free stream velocity.

Fig. 119.5 shows the local friction factor and Stanton number variations for the same case as the contour plots. The friction factor is shown over the whole length of the region but the Stanton number has meaning over the heated patch only. Negative friction factors arise because of the flow reversal, and the flow separation and reattachment positions are indicated at the positions of zero friction factor. The local acceleration with upflow causes a considerable enhancement of the wall shear.

Fig. 119.6 shows mean Nusselt number versus Grashof number for a range of Reynolds number for two of the configurations. Horizontal flows are omitted for clarity and they lie between the curves for the other cases.

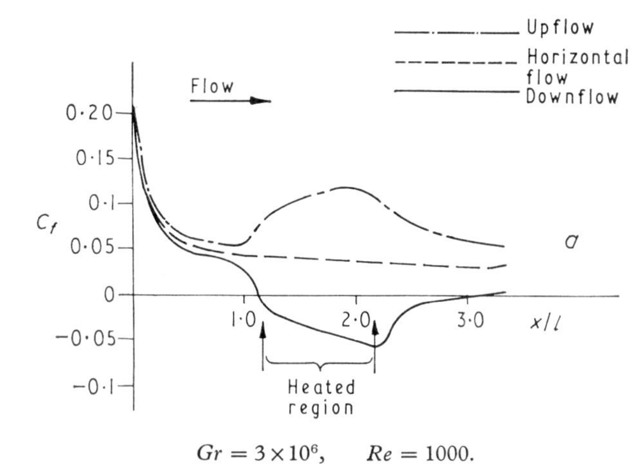

Fig. 119.5a. Friction factor distribution along wall

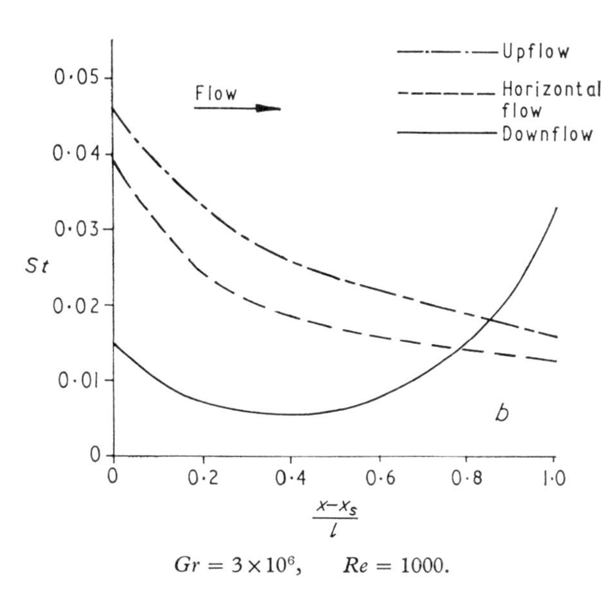

Fig. 119.5b. Stanton number distribution along heated patch

Except at extremely low Reynolds number, the upflow curves do not exhibit any unusual features and the Nusselt number increases with Grashof number. At the very low Reynolds numbers upstream, conduction is significant and warm fluid is carried back over the heated portion thus reducing the wall temperature gradient. The downflow curves show a minimum Nusselt number in all cases as the influence of the natural convection overcomes the forced flow. It is not easy to locate exactly the onset of reversed flow unless a very fine mesh is used, but the region in which this occurs is marked as a shaded zone on the curves. The heat transfer continues to fall as the Grashof number increases after the flow reversal has occurred. This is due to the growth of the recirculating eddy which results in wider spacing of the isotherms close to the wall. Eventually the increasing velocity of the recirculating flow causes the heat transfer to rise.

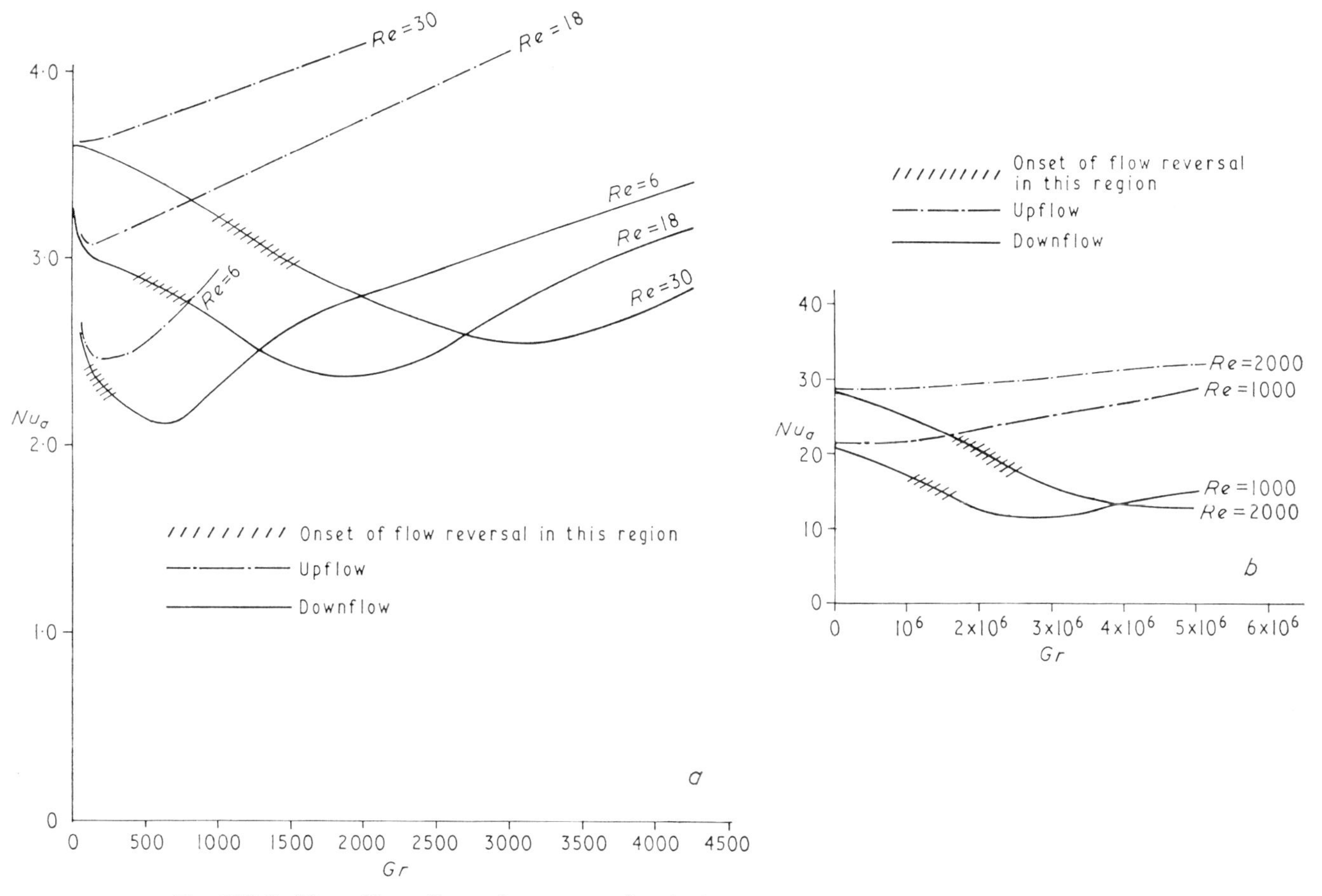

Fig. 119.6. Mean Nusselt number versus Grashof number for different Reynolds numbers

The onset of flow reversal occurs, at the lower Reynolds numbers, when $Gr/Re^2 \sim 2{\cdot}0$ and at the highest Reynolds number when $Gr/Re^2 \sim 0{\cdot}5$.

CONCLUSIONS

Calculations have been made on laminar combined forced and natural convection for a situation in which flow reversal may occur. Whether a steady two-dimensional flow with a recirculating eddy caused by natural convection can exist in reality is open to question. It is well known that inflexions in the velocity profiles give rise to instability.

However, the results should be useful as a guide to the onset of flow reversal with downward flow. The influence of the buoyancy effects on heat transfer and skin friction is considerable in laminar flows. The Nusselt number for downflow may be reduced to only 30 per cent of the corresponding forced flow value.

If a model for the turbulence structure in natural convection, say of the eddy diffusivity type, were to appear it would not be difficult to modify the computer program to include this effect.

APPENDIX 119.1

REFERENCES

(1) PATANKER, S. V. and SPALDING, D. B. *Heat and mass transfer in boundary layers* 1970, 2nd edit. (Intertext Books, London).

(2) GOSMAN, A. D., PUN, W. M., RUNCHAL, A. K., SPALDING, D. B. and WOLFSHTEIN, M. *Heat and mass transfer in recirculating flows* 1969 (Academic Press, London and New York).

(3) RAO, T. L. S. and MORRIS, W. D. 'Superimposed laminar forced and free convection between vertical parallel plates when one plate is uniformly heated and the other is thermally insulated', *Thermodynamics and Fluid Mechanics Convention*, *Proc. Instn mech. Engrs* 1967–68 **182** (Pt 3H), 374.

(4) ACRIVOS, A. 'Combined laminar free and forced convection heat transfer in external flows', *A.I.Ch.E. Jl* 1958 **4** (No. 3), 285.

(5) GUNNESS, R. C. jun. and GEBHART, B. 'Combined forced and natural convection flow for the wedge geometry', *Int. J. Heat Mass Transfer* 1965 **8** (No. 1, January), 1965.

(6) BROWN, C. K. and GAUVIN, W. H. 'Combined free and forced convection—Pt 1. Heat transfer in aiding flow', *Can. J. chem. Engng* 1956 **43** (6), 306; *also* Pt 2, 1956 **43** (6), 313.

(7) HATTON, A. P., JAMES, D. D. and SWIRE, W. H. 'Combined forced and natural convection with low speed air flow over horizontal cylinders', *J. Fluid Mech.* 1970 **42** (Pt 1), 17.

C120/71 A NEW METHOD OF COMFORT HEATING WITH WARM AIR

A. T. HOWARTH* A. S. MORTON† A. F. C. SHERRATT‡

This paper describes a heating system which produces a fairly uniform temperature distribution in an occupied room. An exact inclination of the jet is not critical, thus allowing for installation tolerances. It is shown that the jet can be directed at greater angles away from the floor and still become attached to it.

INTRODUCTION

HEATING SYSTEMS FOR ROOMS usually require the heat supplied to be equal to that lost by heat transfer through the room surfaces. Infiltration and solar gain are important factors which also influence the heating requirements. In winter, with lower outside air temperatures, there can be large temperature gradients across external walls, windows, etc., and temperature gradients within the room air, particularly in the region of these boundaries. It is to be expected that air movement will be generated in these regions by a process of natural convection.

Present warm air heating systems tend to produce large floor to ceiling temperature gradients of the order of 5–8 degC. Gradients of this magnitude do not produce comfortable environments for living (1)§. This paper describes tests on a warm air system which is potentially capable of reducing or eliminating these large temperature variations. The air is introduced into the room through a long slot running the length of one wall. It issues from the slot in a direction that can be parallel to the floor or slightly inclined in an upwards direction, forming a carpet of air travelling across the floor.

The success of the system is attributable to the following:

(*a*) Turbulent jets issuing from long narrow slots induce entrainment and mixing of the air in the room with that in the jet.

(*b*) The effect of natural convection, which often causes heated jets to form an upward curving trajectory, is counteracted by the Coanda effect which is stronger when the jet does not have open ends through which surrounding air may be drawn (Fig. 120.1).

Notation

- T_{am} Average of the air temperature readings as measured by the thermocouples in the locations shown in Fig. 120.5.
- T_{cc} Temperature in the cold corridor (west wall).
- T_{cm} Average inside ceiling surface temperature.
- T_g Average inside surface temperature of the window glass.
- T_j Jet temperature at the slot outlet.
- $T_j - T_g$ Difference between jet temperature and minimum inside surface temperature of the window glass.
- T_{oc} Temperature in cavity of ceiling.
- T_{oe} Temperature in cavity of east wall.
- T_{on} Temperature in cavity of north wall.
- T_{os} Temperature in cavity of south wall.
- T_{wm} Average of the wall inside surface temperatures (excluding the window surface temperatures).
- ΔT_a Difference between the maximum air temperature at a height of 7·5 ft (2·29 m) and the minimum air temperature at 0·5 ft (0·15 m).
- V Jet volume flow rate.
- θ Angle of elevation of slot axis floor.

EXPERIMENTAL TECHNIQUES

The tests were conducted in a special room (2)–(4) with hollow walls and dimensions 5·5 m × 3·68 m × 2·44 m (height). Two adjacent walls were transparent, the other walls being opaque with a matt black interior finish.

The variables which govern the conductive heat loss from rooms are the temperatures in the adjacent zones and the thermal transmittance of the walls. In the test room used for this work, a cold region, simulating the winter outdoor environment, was constructed by forming a refrigerated corridor adjacent to the west wall of the room. The simulation of neighbouring indoor regions was

The MS. of this paper was received at the Institution on 18th May 1971 and accepted for publication on 11th June 1971. 33

The research described in this paper was carried out in the Building Science Laboratories, Dept of Architecture, University of Nottingham. Certain aspects of the invention described in this paper, and its application, are covered by British Provisional Patent No. 56422/70 (National Research and Development Corporation).

* *Dept of Construction, Oxford Polytechnic.*
† *Dept of Mechanical Engineering, University of Nottingham.*
‡ *Dept of Architecture, University of Nottingham.*

§ *References are given in Appendix 120.1.*

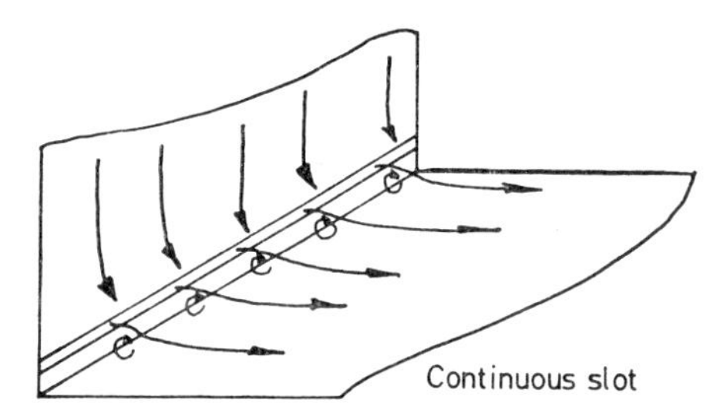

Fig. 120.1. Strong Coanda effect

effected by providing horizontally located electrical heating wires in the cavities of the walls and ceiling. Observation of the air movement in the room was by a visualization technique described by Daws (5). Visible white particles* which had a low terminal velocity were injected into the room, and vertical planes were highlighted using a light source producing a vertical parallel beam projected through the transparent ceiling. The flow in mutually perpendicular planes of interest, each plane being parallel to one or other of the room axes, was observed through the appropriate transparent wall. Quantitative measurements were obtained by photographing the moving particles during an exposure time of 0·25, 0·5, or 1·0 s. The matt black walls provided the contrasting background that is required for the photography of the white particles. (Figs 120.5 and 120.6 show typical photographs.)

Temperatures were measured using 36 s.w.g. copper–constantan thermocouples. Twelve of these thermocouples were mounted on a moving grid for the purpose of measuring the air temperature (Fig. 120.2). A further 21 thermocouples sensed the temperatures at various points on the interior surfaces of the room (Fig. 120.3), and mercury-in-glass thermometers indicated the temperatures in the wall and ceiling cavities.

* *The particles were produced by heating metaldehyde tablets, forming 'lightweight' crystalline particles, of up to 6 mm maximum diameter, which had the appearance of snowflakes.*

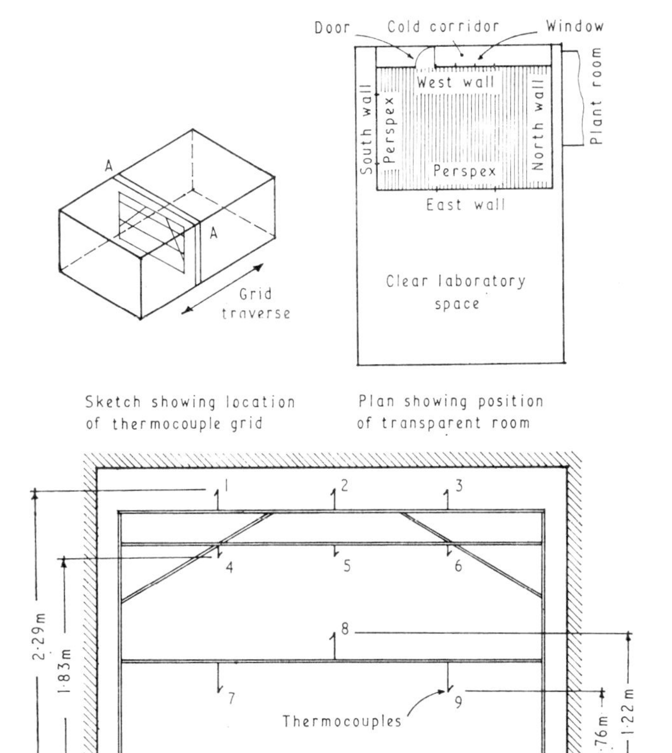

Fig. 120.2

The air was supplied from a fan, via an electric heater battery and ducting in which a flow measuring orifice plate was situated, to three adjacent slot units skirting the east wall. The air, after circulating through the room, was returned to the fan through an outlet grill located on the north wall. The results of previous workers (5) (6) suggested that the location for the return grill should not

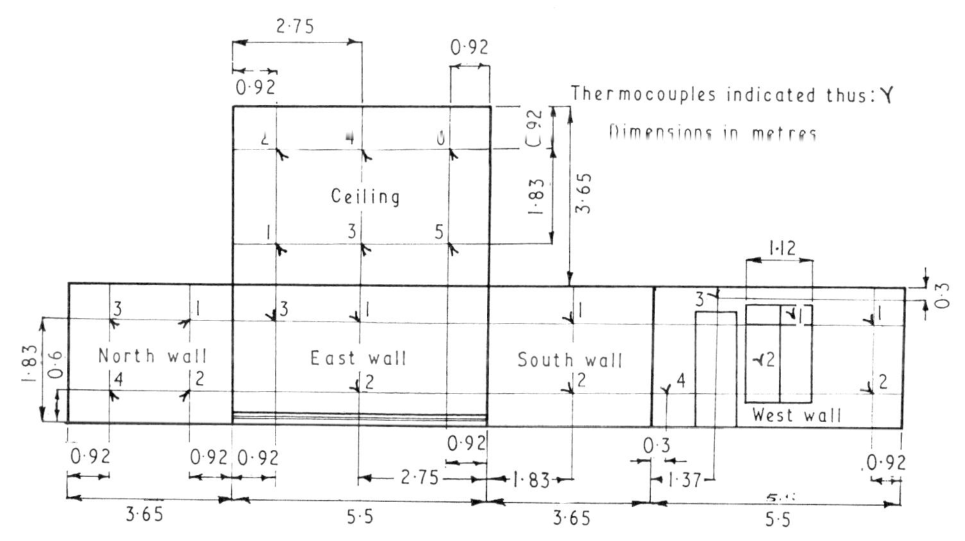

Fig. 120.3. Orthographic view of room interior showing thermocouples

affect the air movement pattern in the room to any appreciable extent. This was confirmed during the course of the present work which showed that the air approached the return grill from all directions and the effect on the overall motion of the room air takes the form of a gradual drift, where the drift velocities are small compared with those concerned with the main air currents in the room, which are substantially two-dimensional. Using the photographic technique assessment of velocities is considerably simplified where the flow is two-dimensional in the beam of light because the particle trace on the photograph then represents the actual distances moved in the exposure time.

THE DESIGN OF THE LONG SLOT

The slot sections (Fig. 120.4) designed for the present work had a slot width of 6·5 mm, the slot being divided into three sections each of 1·83 m length and having its own inlet to fit at the base of the 5·5-m east wall. In order to ensure that the air discharge was perpendicular to the slot face, strips of corrugated material were fitted into the slot so that, effecively, the air was being discharged through a series of tubes approximately 5 mm diameter and 4·5 cm long, closely packed together. After allowing for the cross-sectional area of the corrugated material, the ratio of slot area to plenum area for each section is 0·222.

The angle of elevation of the jet discharge was varied by swivelling the units as a whole.

THE EXPERIMENTAL INVESTIGATION

In studying the air movement and temperature distribution within the room the effect of the following variables was investigated:

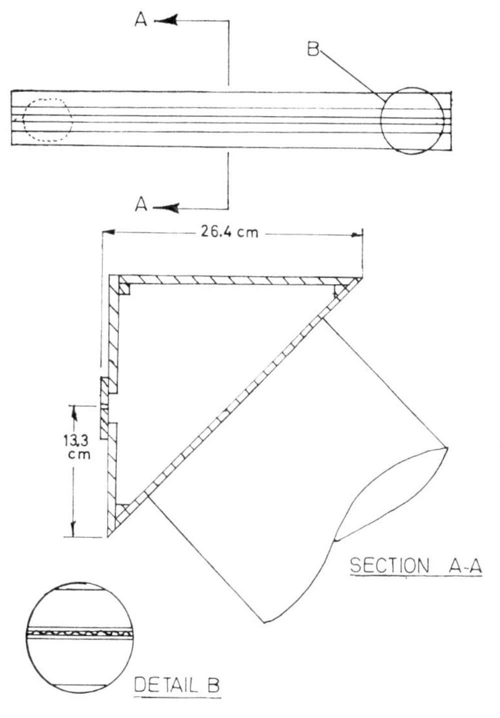

Fig. 120.4. Slot–plenum design

(*a*) air jet temperature (T_j);
(*b*) air flow rate through the slot (V);
(*c*) jet discharge angle of elevation (θ);
(*d*) temperatures in wall cavities and cold corridor.

The tests were conducted in two groups: the first with the jet temperature equal to the room air temperature (i.e. isothermal), and the second with jet temperatures of around 50°C. During the tests with warm air the temperature in the cold chamber was controlled at 5°C. The wall heaters were used in such a way that for isothermal tests all the temperatures in the cavities were maintained at approximately the level of the warmest cavity during shutdown conditions. During the isothermal tests the refrigeration plant in the cold chamber was not used. For warm air tests the cavities were maintained at a temperature that was slightly higher than that in the laboratory. Four flow rates of 5, 7·5, 10, and 15 air changes/h were used for each of three discharge angles of the air jet ($\theta = 0°, 30°, 42°$ or $45°$) in both the heated and unheated cases.

The air velocities in the room were measured using the techniques described previously. Since the air handling equipment was located outside the main laboratory area and the tests were conducted during the winter months, it was found necessary to apply a rate of heating of 1 kW to the air that was recirculated during the isothermal tests. Temperatures were only recorded under steady conditions at 5 air changes/h as it was assumed that the increased mixing associated with the higher flow rates produced a better temperature distribution within the room.

Periodic checks were made of the two-dimensionality of the flow by observing and photographing a plane of the room normal to the initial jet direction.

PRESENTATION OF THE RESULTS

Figs 120.5 and 120.6 show a summary of the results that were recorded. The symbolism used in these diagrams is described in the Notation and temperatures are given in Tables 120.1 and 120.2. The temperatures used in the diagrams ($\theta = 0°$ in Fig. 120.5 and $\theta = 0°$ and 30° in Fig. 120.6) are representative values to indicate the magnitudes of the temperatures of the air in the jet itself and of the occupied zone, walls, and ceiling. ΔT_a denotes the difference between maximum temperature at a height of 2·3 m and minimum at a height of 0·15 m above the floor, and should therefore be representative of the maximum temperature variation occurring in the occupied zone away from the floor. In the tests using warm air the cavity temperatures are denoted by the symbols T_{oe}, T_{os}, and T_{on}, depending on whether the wall in question was the east, south, or north wall (see Fig. 120.2). The sketches indicate the flow directions.

DISCUSSION OF THE RESULTS

General observations

It is clear that the Coanda effect is present when the angle of elevation (θ) of the jet is less than 42°, because the entering air stream tends to move towards the floor, and in

Table 120.1

Surface temperatures during isothermal flow (°C)
5 air changes/h (see Fig. 120.3)

Surface	Thermo-couple no.	$\theta = 0°$	$\theta = 30°$	$\theta = 42°$
Jet		20·8	20·8	21·3
Window	1	20·4	21·4	21·8
	2	20·3	21·4	21·4
South wall	1	20·9	20·8	21·5
	2	20·3	20·4	21·3
North wall	1	21·0	20·5	20·2
	2	20·3	20·1	18·9
	3	20·8	20·3	19·6
	4	20·4	20·1	18·8
East wall	1	21·3	20·8	21·4
	2	20·5	20·5	21·5
	3	21·0	20·8	21·8
West wall	1	20·5	20·3	20·0
	2	20·1	20·3	20·0
	3	21·5	21·0	21·3
	4	20·5	20·8	20·9
Ceiling	1	21·5	21·2	21·7
	2	21·0	20·9	21·7
	3	21·3	20·1	20·3
	4	21·3	21·0	21·9
	5	21·2	21·0	21·9
	6	21·3	21·1	21·9

Air temperatures (°C)
5 air changes/h (see Fig. 120.2)

Thermocouple no.	$\theta = 0°$	$\theta = 30°$	$\theta = 42°$
R1	21·1	20·1	21·1
R2	21·3	20·1	21·1
R3	21·3	20·1	21·1
R4	20·8	20·1	21·1
R5	20·6	20·1	21·1
R6	20·7	20·1	21·1
R7	20·5	20·1	21·1
R8	20·3	20·1	21·1
R9	20·3	20·1	21·1
R10	20·0	20·1	20·8
R11	20·3	20·0	20·8
R12	20·4	19·9	20·1

Table 120.2

Surface temperatures during flow of warm air (°C)
5 air changes/h (see Fig. 120.3)

Surface	Thermo-couple no.	$\theta = 0°$	$\theta = 30°$
Jet		51·0	49·4
Window	1	15·2	14·4
	2	17·1	17·1
South wall	1	21·0	21·8
	2	20·4	21·4
North wall	1	22·0	25·9
	2	23·0	25·9
	3	22·0	25·9
	4	22·5	26·4
East wall	1	21·2	22·2
	2	20·8	22·2
	3	20·9	22·2
West wall	1	22·0	23·5
	2	22·0	23·6
	3	22·0	23·7
	4	21·4	22·5
Ceiling	1	21·4	23·1
	2	21·4	23·7
	3	23·6	25·2
	4	21·4	22·7
	5	21·5	23·5
	6	21·4	23·5

Air temperatures (°C)
5 air changes/h (see Fig. 120.2)

Thermocouple no.	$\theta = 0°$	$\theta = 30°$
R1	28·9	28·4
R2	28·9	28·4
R3	28·9	27·7
R4	28·9	28·6
R5	28·6	27·7
R6	28·6	27·7
R7	28·6	27·7
R8	28·6	27·9
R9	28·6	27·9
R10	28·6	26·5
R11	28·6	28·1
R12	28·6	29·7

most cases adheres to it, travelling directly across the room.

With jet angles of elevation of less than 42° the most noticeable movement in the room during all the tests is that of the downward flowing air stream caused by entrainment near the jet outlet.

The air temperature variation, ΔT_a, is always less than 1·3 degC. The value of ΔT_a, by definition above, excludes the higher temperatures which may be present in the proximity of the air inlet (i.e. thermocouple No. 12), but it is interesting to note that only in the heated case, with a jet angle of $\theta = 30°$, does this temperature significantly exceed that recorded at thermocouples 10 and 11 (Table 120.2). This is to be expected because the heated jet issuing in an upward direction is more likely to affect the lower thermocouple readings than the jets travelling parallel to the floor.

Although a full analysis of the velocity distribution in the jet travelling (in most cases) along the floor was not carried out, a simple calculation—assuming a slot area of 175 cm²—indicated that for an air change rate of 5 per hour the mean air velocity at the jet outlet was 3·93 m/s. This estimated velocity increases in proportion to the air flow rate, but it was found that even at the higher flow rates used, the air movement was not prominent subjectively more than 0·45 m away from the outlet. Problems of discomfort due to these outlet velocities are not likely to occur because the legs and feet tend to be less sensitive to air currents than the head and neck.

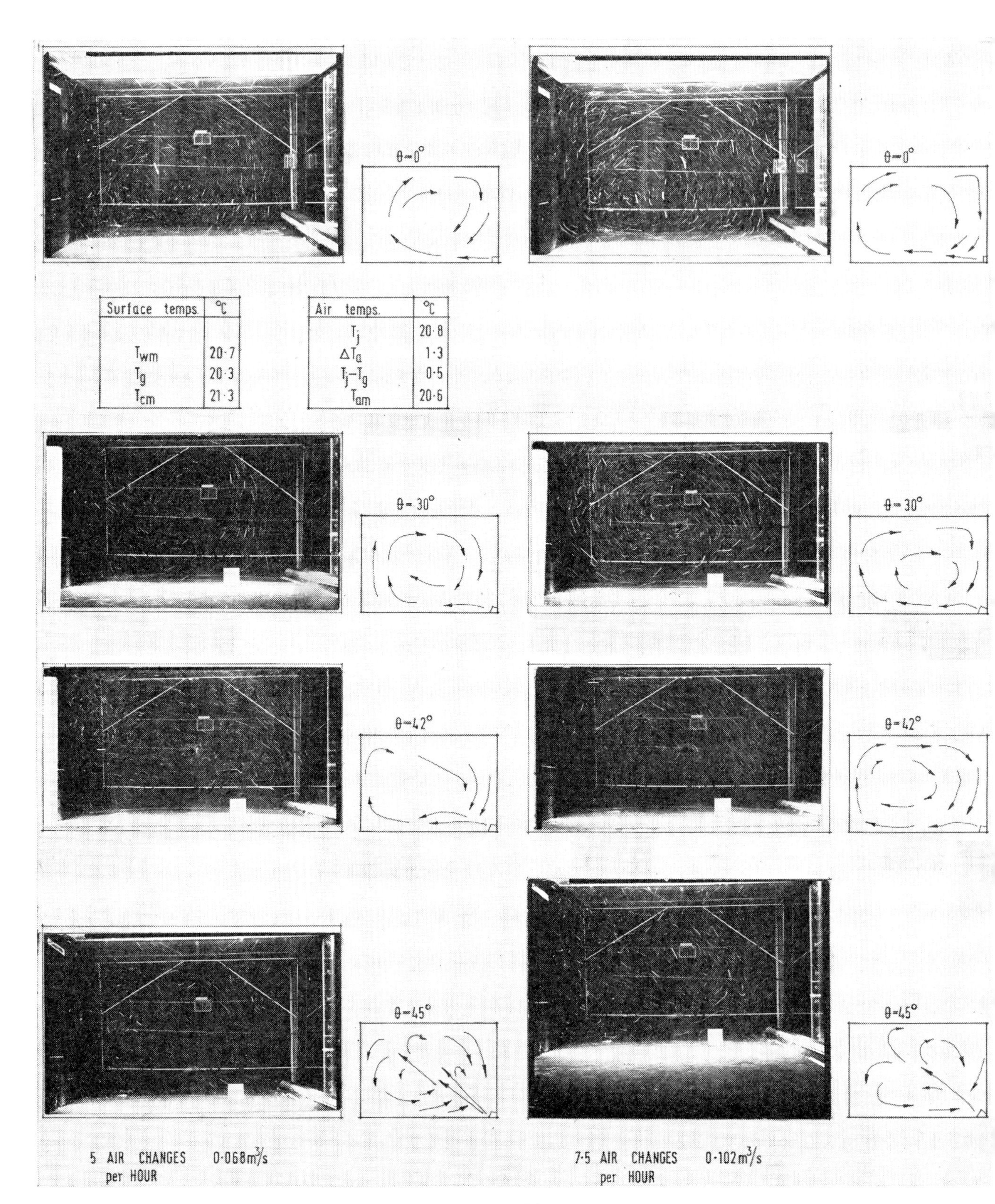
θ = 0°
θ = 0°
Surface temps. °C
T_{wm} 20·7
T_g 20·3
T_{cm} 21·3
Air temps. °C
T_j 20·8
ΔT_a 1·3
$T_j - T_g$ 0·5
T_{am} 20·6
θ = 30°
θ = 30°
θ = 42°
θ = 42°
θ = 45°
θ = 45°
5 AIR CHANGES per HOUR 0·068 m³/s
7·5 AIR CHANGES per HOUR 0·102 m³/s

Fig. 120.5

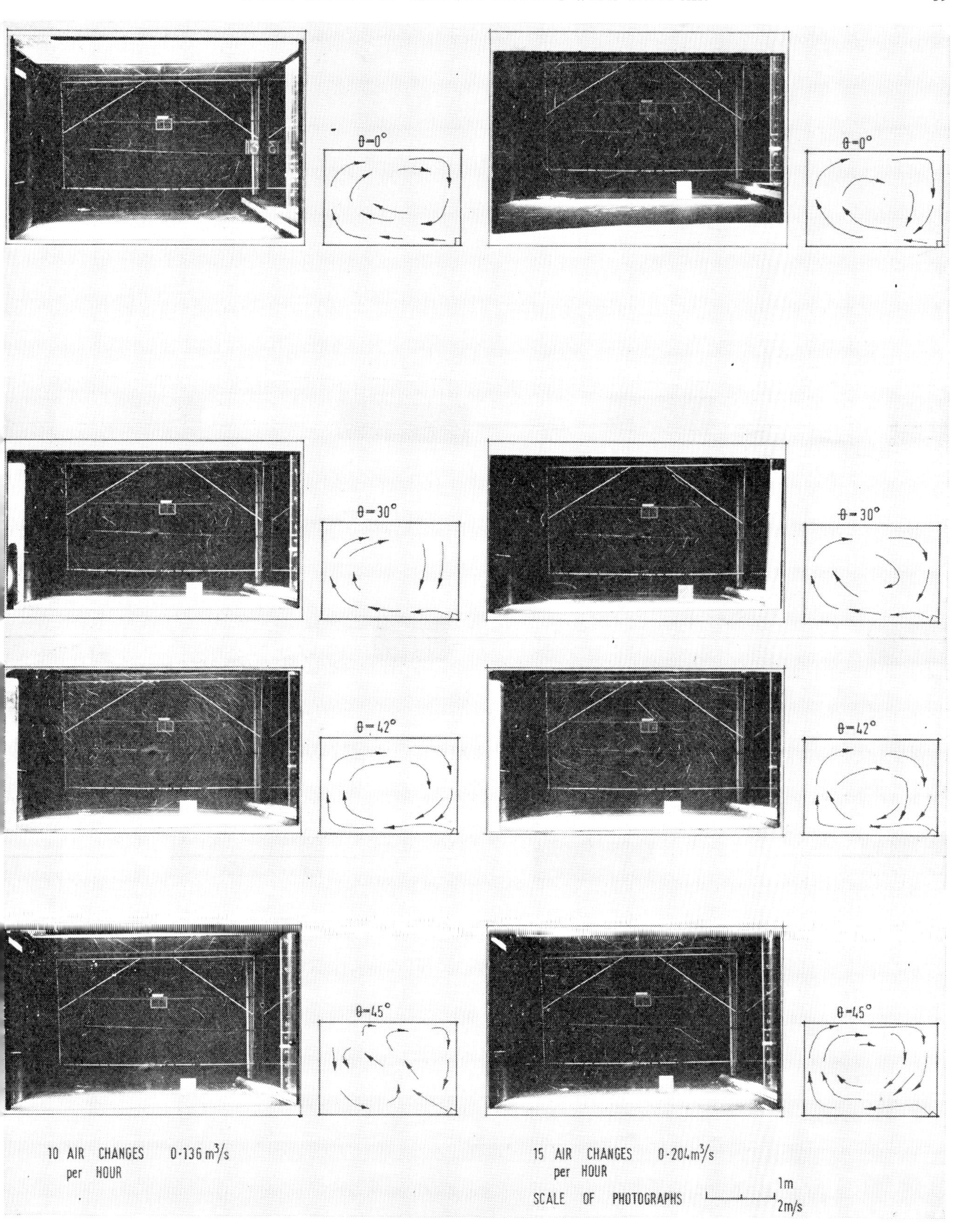

Fig. 120.5—contd

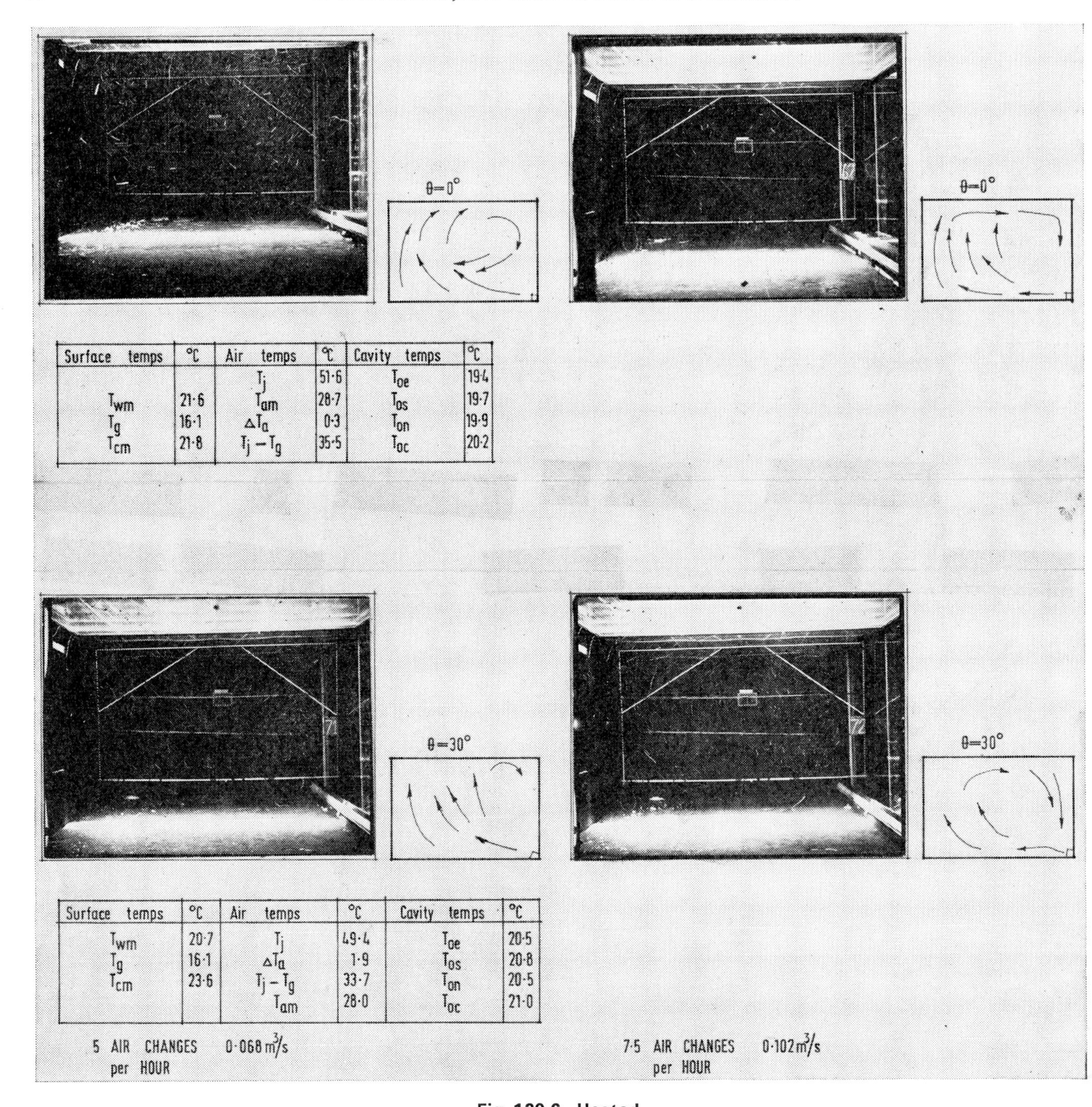

Surface temps	°C	Air temps	°C	Cavity temps	°C
		T_j	51·6	T_{oe}	19·4
T_{wm}	21·6	T_{am}	28·7	T_{os}	19·7
T_g	16·1	ΔT_a	0·3	T_{on}	19·9
T_{cm}	21·8	$T_j - T_g$	35·5	T_{oc}	20·2

Surface temps	°C	Air temps	°C	Cavity temps	°C
T_{wm}	20·7	T_j	49·4	T_{oe}	20·5
T_g	16·1	ΔT_a	1·9	T_{os}	20·8
T_{cm}	23·6	$T_j - T_g$	33·7	T_{on}	20·5
		T_{am}	28·0	T_{oc}	21·0

Fig. 120.6. Heated

Detailed observations

The isothermal case

Fig. 120.5 shows the effect of progressively increasing the jet discharge at various angles of elevation. For $\theta = 0°$ and 30° it can be seen that a circulatory type of flow is established in which the general level of velocities increases with jet outlet velocity. If we define stagnant regions as those where the velocity is less than 0·1 m/s it can be seen that they are small in area and, in general, are near the centre of the room, and it was found on occasions that with the higher jet velocities other flow patterns are established in which stagnant regions occurred in other parts of the room, although the flow continued to be circulatory.

It would appear from these results of isothermal tests that in order to create a general air movement pattern in the room with desirably small stagnant zones, nothing is gained by increasing the flow rate above 5 air changes/h. When the jet angle of elevation was increased to 42°, some instability of flow was noticed in which the jet would periodically break away from the floor and flow diagonally across the room for a few seconds and then re-attach to the floor. Fig. 120.5, at $\theta = 42°$, shows the flow at instants

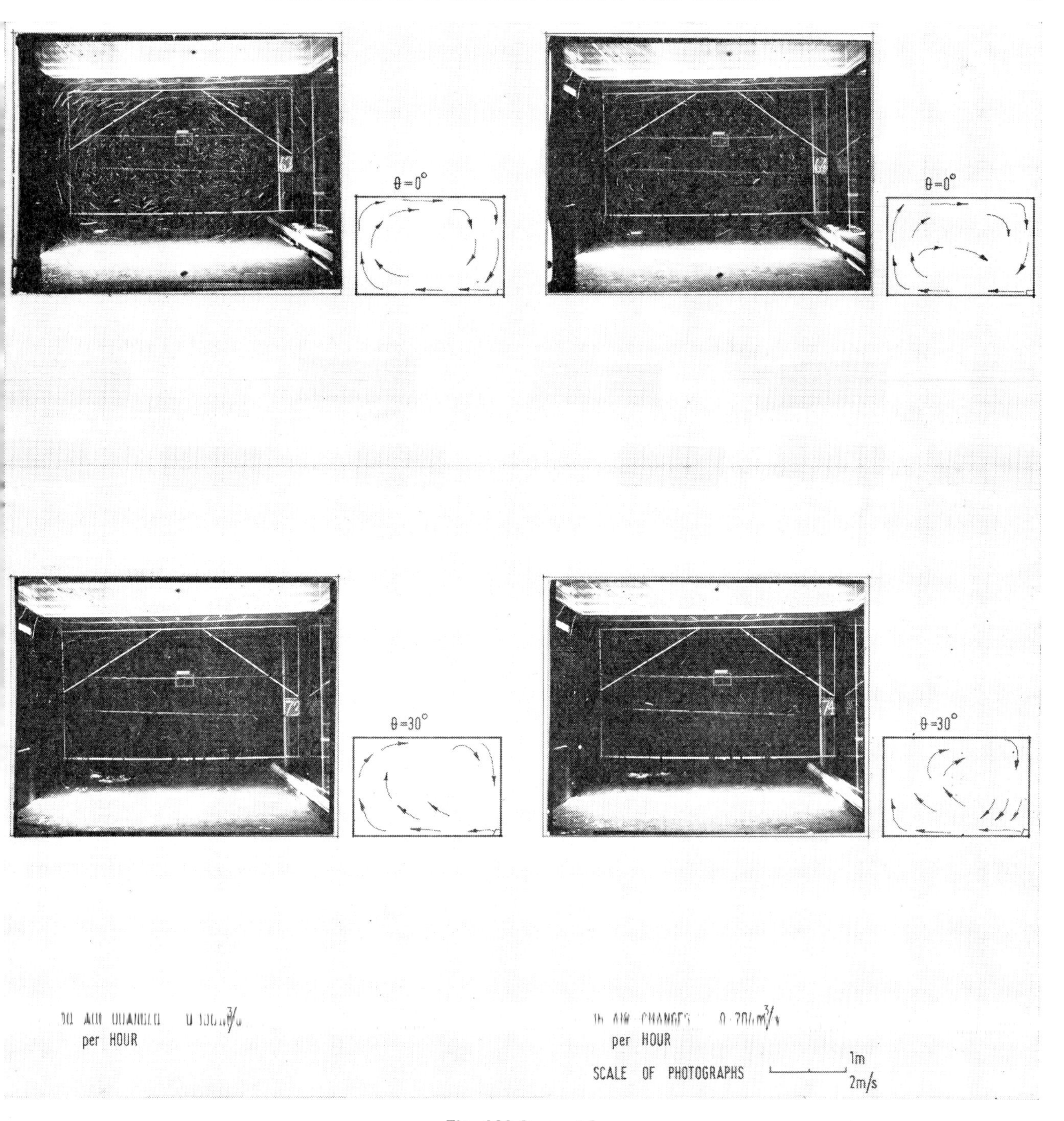

Fig. 120.6—contd

when the jet was in the attached condition. When the jet angle of elevation was further increased to 45° the flow was generally in a diagonal direction.

The heated case

The temperatures shown in Fig. 120.6 are derived from those recorded in Table 120.2. T_{wm}, T_g, and T_{cm} are representative values indicating the magnitudes of the temperatures at the inside surfaces of the walls, window, and ceiling respectively. T_{am} is the mean air temperature in the zone of the room which is most likely to be occupied, and excludes the lower 0·15 m. The maximum temperature difference in the room exists between the slot face and the window surface and is represented by $(T_j - T_g)$. Despite a value of 35·8 degC for this latter temperature difference, the maximum temperature difference recorded in the occupied zone was only 0·3 degC for $\theta = 0°$ and 1·9 degC

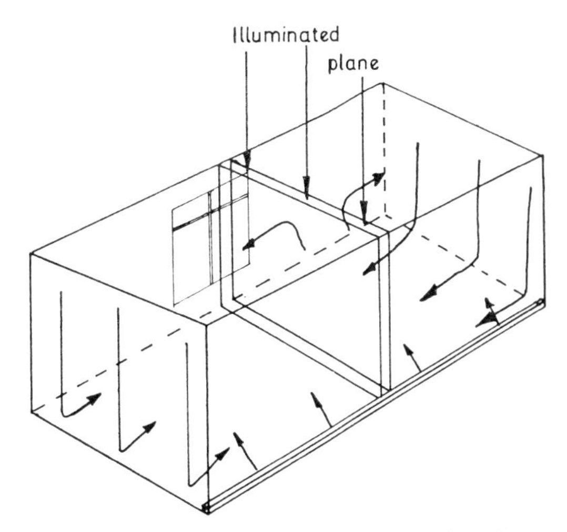

Fig. 120.7. Descent of air near cool walls

for $\theta = 30°$, indicating the thoroughness of the mixing of the warm and cool air.

For the jet velocities and angles of elevation in which two-dimensional flows occurred under isothermal conditions, the presence of comparatively cool end walls brings about a lack of two-dimensionality in the flow which is accentuated by the localized cold area of glass at the window. The effect of these cool surfaces on the air movement depends on the flow rate of incoming air (Fig. 120.6, $\theta = 0°$). At the lower rates of 5 and 7·5 air changes/h the air stream rises before reaching the opposite wall and flows towards the ceiling. The air then moves into a plane at right angles to the plane of the photograph and flows towards the north and south end walls. Previous work (2) (7) suggests that in the room without ventilation, the flow rate at the base of a 2·44 m high wall, which is 7 degC cooler than the room air (T_m), is 0·008 m^3/s per metre width of wall. The naturally convective flow rate at the base of each of the north and south walls could be about 0·03 m^3/s (7). This relatively cool volume of air tends to fall down the walls until it meets the floor, along which it travels in the plane perpendicular to the plane of the light beam (Fig. 120.7). Considering both end walls, a volume of around 0·06 m^3/s is involved in this natural convection, and this is the same order of magnitude as the amount of warm air which is forced into the room with air change rates of 5 (0·068 m^3/s) and 7·5 (0·102 m^3/s) per hour. Hence, the air movement forms a three-dimensional flow distribution which is not fully represented by the flow pattern on the room axis indicated in Fig. 120.6—$\theta = 0°$, 5 and 7·5 air changes/h.

At the higher flow rates it appears that the heat exchange within the room is predominantly by forced convection. It can be seen that the motion of the air is similar to that of the isothermal studies—compare Figs 120.5 and 120.6 (10 and 15 air changes/h) when the jet angle is 0°. Clearly the incoming warm air swamps the naturally convective motion described above. This similarity between the higher flow rates of the heated and isothermal jets at $\theta = 0°$ does not exist when $\theta = 30°$. The effect of natural convection, combined with the initial upward inclination of jet direction, seems to be sufficient to reduce the Coanda effect even at air change rates of 10 and 15 per hour (Fig. 120.6, $\theta = 30°$). When θ was further increased to 45°, the instabilities observed were similar to those in the isothermal case.

The effect of occupants and obstructions

Large obstructions such as tables, settees, sideboards, etc., must inevitably cause drastic modifications of the air patterns. However, unless the obstruction completely blocks the path of the jet (e.g. furniture beneath which there is a very small clearance) it is to be expected that the Coanda effect will remain, even though the room air currents might be changed. Hence a reasonable temperature distribution should be maintained.

CONCLUSIONS

A tested system for providing room heating has been shown to produce fairly uniform temperature distributions in the occupied room with velocities that are not likely to produce discomfort. Furthermore, the exact inclination of the jet is not critical, thus allowing for installation tolerances. In fact it has been shown that the jet can be directed at angles away from the floor much in excess of those likely to be incurred and still become attached to the floor.

ACKNOWLEDGEMENTS

The authors thank all those who assisted in this work, particularly the useful discussions with Mr L. F. Daws of the Building Research Station and the technical assistance given unstintingly by Mr C. H. Riley, Building Science Laboratories, University of Nottingham. The financial support, by way of a Science Research Council grant to Dr A. F. C. Sherratt, is gratefully acknowledged.

APPENDIX 120.1

REFERENCES

(1) BRUCE, W. 'Man and his thermal environment', *N.R.C. Tech. Paper 84*, 1960.

(2) HOWARTH, A. T. 'An investigation of factors affecting air movement in mechanically ventilated rooms', M.Phil. thesis, University of Nottingham, 1970.

(3) SHERRATT, A. F. C., HOWARTH, A. T. and MORTON, A. S. 'Facility for investigating room air movement', Paper presented at the Fifth International Congress for Heating, Ventilating and Air Conditioning, Copenhagen, 1971 (May).

(4) HOWARTH, A. T., SHERRATT, A. F. C. and MORTON, A. S. 'The design and construction of a test room for heating and ventilating studies' (in preparation).

(5) DAWS, L. F. 'Movement of air streams indoors', Building Research Station Research Paper 66, 1967 (August) (Building Research Station, Garston, Watford, Herts.).

(6) STRAUB, H. E., GILMAN, S. F. and KONZO, S. 'Distribution of air within a room for year round air conditioning—Part 1', *Bull. Ill. Univ. Engng Exp. Stn* 1956 (No. 435).

(7) HOWARTH, A. T., SHERRATT, A. F. C. and MORTON, A. S. 'Air movement in an enclosure with a single heated wall', *J. Instn Heat. Vent. Engrs* (in the press).

Discussion

P. H. G. Allen Member

In spite of elegant analyses for developing flow conditions in simple geometries (see, e.g., Papers C115/71 and C116/71 in this Symposium), the concept of fully developed velocity and temperature profiles remains extremely useful when designing complicated systems such as transformer windings (**1**). I wish to comment on the buoyancy term in the momentum equation governing fully developed, non-isothermal, laminar flow in vertical ducts. This term represents a force acting on each fluid element in the flow cross-section. It is proportional to the difference in fluid density between the element and a certain datum. In Paper C112/71 one can deduce from equation (112.1) that the datum chosen is at the temperature of the outer wall of the annulus. In the corresponding equation of Savkar's paper (**2**), the datum is evidently at the local bulk temperature of the fluid. Other authors present a variety of alternatives. I believe the correct datum density is at the arithmetic mean temperature of the fluid in the section considered.

It is well known that, by comparing the pressures exerted by equal heights of fluid in the two vertical portions of a closed convective system and then integrating for all heights, the unbalanced pressure head causing flow is given by:

$$\left(g\frac{d\rho}{dt}\right)\times(\text{area within height}-\text{temperature diagram for system})$$

(with notation as in Paper C112/71), where ρ varies linearly with t. This simple analysis assumes that in every horizontal cross-section in the flow path temperature, and hence fluid density, is uniform over the section. However, with laminar flow heat transfer this is far from true and we must evaluate from first principles the element of pressure Δp_ρ due to an elementary horizontal layer of non-isothermal fluid, thickness Δx. By definition:

$$\Delta p_\rho=\frac{\text{force}}{\text{area}}=\frac{\text{integral of }(g\rho\,\Delta x)\text{ over duct cross section}}{\text{duct cross-section area}}$$

$$=g\rho_a\,\Delta x \quad . \quad . \quad . \quad . \quad . \quad . \quad . \quad . \quad . \quad (1)$$

where ρ_a is the arithmetic mean density for the section. In the limit:

$$\left(\frac{\Delta p_\rho}{\Delta x}\right)_{\Delta x\to 0}\to\frac{dp_\rho}{dx}=g\rho_a \quad . \quad . \quad . \quad (2)$$

and thus it is evidently the density at the arithmetic mean temperature around the flow path that must be considered when calculating convective pressure head.

The pressure gradient term in the momentum equation is usually written:

$$\frac{dp}{dx}-g\rho$$

with p defined simply as pressure. This difference represents the gradient of p', the pressure causing flow. We may write:

$$\rho=\rho_a+(\rho-\rho_a) \quad . \quad . \quad . \quad . \quad (3)$$

so that the pressure gradient term becomes:

$$\frac{dp}{dx}-g\rho_a-g(\rho-\rho_a)$$

Each constituent may be identified:

(i) dp/dx: this integrates around the flow path to give the component of pressure head due to pump action, if any.

(ii) $g\rho_a$: this integrates around the closed flow path to give the component of pressure head due to natural convection.

(i) and (ii) add to give the gradient of p', which is uniform over the duct section.

(iii) $g(\rho-\rho_a)$: this varies over the duct section and distorts the velocity pattern compared with isothermal.

Written in general vector notation, the Navier (momentum) equation is thus:

$$\text{div}(\mu\,\text{grad}\,u)=\frac{dp'}{dx}-g(\rho-\rho_a) \quad . \quad . \quad (4)$$

This treatment follows that due to Ostrach (**3**) except that he defined what I have called Δp_ρ on a static, isothermal, basis and then applied it to a nonisothermal situation. By taking his datum density value ρ_s as the hydrostatic one, he obscured the true (arithmetic mean) datum value.

I consider that my own treatment clarifies the nature of mixed convection by differentiating between the two aspects of natural convection, namely:

(*a*) Generating a pressure head which in mixed convection combines with that due to pump action.

(*b*) Perturbing the velocity distribution.

The importance of taking the correct datum density as

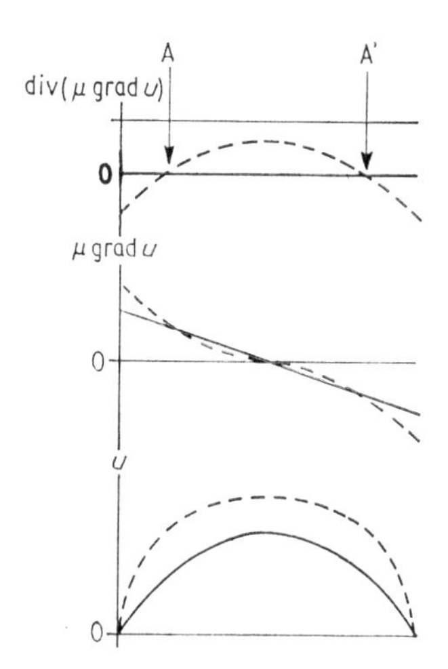

Fig. D1. Effect of convective perturbation (dashed lines) compared with isothermal (full lines) on successive integration to give fully developed velocity (u) distribution

opposed to an arbitrary one can be seen from the common case of density varying linearly with temperature. Then, for equation (3) we may write:

$$\rho = \rho_a + \frac{d\rho}{dt}(t - t_a) \quad . \quad . \quad . \quad (3a)$$

so that equation (4) becomes:

$$\text{div}(\mu \text{ grad } u) = \frac{dp'}{dx} - g\frac{d\rho}{dt}(t - t_a) \quad . \quad (4a)$$

Remembering that dp'/dx, $d\rho/dt$ and (in upward flow) g are all negative quantities, evidently the second term on the right-hand side of equation (4a) reinforces the first for $t > t_a$ and opposes it for $t_a > t$. Thus the datum (here t_a) is critical in determining the quantitative nature of the perturbation of the velocity distribution. Fig. D1 illustrates this qualitatively. Here $t = t_a$ at A and A'. Choosing a different datum would result in a different value of dp'/dx when iterating for a specified u_m and thus invalidate pressure drop calculations.

REFERENCES

(1) ALLEN, P. H. G. and FINN, A. H. 'Transformer winding thermal design by computer', 1969, I.E.E. Conference Publication 51, 589.

(2) SAVKAR, S. D. 'Developing forced and free convective flows between two semi-infinite parallel plates', Paper NC 3.8, *4th Int. Heat Transfer Conf.*, Paris, 1970.

(3) OSTRACH, S. 'Laminar natural-convection flow and heat transfer of fluids with and without heat sources in channels with constant wall temperatures', 1952, N.A.C.A. Technical Note 2863.

F. Porges Member

Paper C117/71: There are assumptions made in the theory which I would not have expected to be valid, and I must admit to some surprise that the test results in fact confirm the theory. However, the theory contains a coefficient of discharge which had to be determined by experiment, and any shortcomings of the theory would show up only as discrepancies in this coefficient. The graph of the coefficient plotted in Fig. 117.4 is a smooth curve, which shows that any discrepancies are systematic and not random. Such a systematic variation of an experimentally determined coefficient is consistent with the use of a theory which has one or two terms missing.

The first assumption I would question is that because the enclosure is sealed there is no net flow of air across the opening. The densities and temperatures in the two rooms are different, and one might quite reasonably expect to find a mass flow across the opening to equalize them.

Mathematically, this assumption is contained in the statement that the absolute pressure is equal at the centre line of the opening. I find it very hard to visualize any physical reason why it should be. It would be interesting to have the actual experimental results, to see where the line of equal pressure in fact was.

If the line of equal pressure is not at the centre, then the limits of integration in equation (117.6) are different. Suppose they are not 0 to $H/2$, but 0 to $aH/2$, where a is an unknown factor. Then equation (117.6) is modified to

$$Q = Ca^{3/2}\frac{W}{3}\left[g\frac{\Delta\rho}{\bar{\rho}}\right]^{1/2} H^{3/2}$$

Since $Ca^{3/2}$ is a constant, and C is found by experiment, it could well be the case that the constant found and called C is in fact the combined constant $Ca^{3/2}$. What I am suggesting is that the observed variation of C is in fact not a variation of C at all, but of another constant with which C has been unknowingly multiplied.

The other assumption made which I would question is that there is no temperature gradient in either room. In any real case there will be a gradient of temperature, and therefore of density, which will modify the initial equations from which everything else is derived. I have not attempted to rewrite them to see what happens, but it may be that the effect is similar to shifting the line of equal pressures, in which case the consequences are the same as those I have already suggested.

There are some grounds for thinking that this is so. In Fig. 117.4 the coefficient of discharge is constant at temperature differentials large enough to make the temperature gradient negligible in comparison. But when the differential between the rooms is of the same order of magnitude as the gradient one would expect in either of them, the coefficient varies, and varies in a systematic manner. I suggest that this is what one might expect of a theory which makes an assumption not valid at all values of the variables considered.

Paper C120/71: In the discussion of results the authors say that for an air change rate of 5/h, the jet outlet velocity was 3·93 m/s, and at the higher rates it was proportionately higher. I have for some time made it a rough rule of design that to eliminate noise in ventilation systems the velocity in ducts within dwellings should not be more than about 4 m/s, and the outlet velocities at grilles should preferably be half this. It seems to me that a jet issuing at 3·93 m/s might be just audible, and I wonder whether the authors feel it might be prudent to try to reduce the jet velocity and whether this could be done without a significant reduction in the Coanda effect.

The authors have very briefly referred to the effect of furniture within a room. One of the reasons for the delight with which architects originally turned to warm air heating was the difficulty they had always had in finding wall space for radiators. The heating engineer who now wants to suggest a grille running the whole length of a wall is going to find his clients very resistant to the idea, and I think it may be the clients and not the heating engineer who are right. Bookcases are solid at floor level, and so are some sideboards and settees. Wardrobes, fitted cupboards, and pianos are also common pieces of furniture which would get in the way of the full length grille.

The system described will also require much more extensive ducts and plenums behind the walls, so that many more voids will have to be provided within the structure.

Any method of heating imposes restraints on the planning and layout of a building. Warm air heating became popular fairly suddenly ten or twelve years ago because it seemed to impose fewer restraints than conventional radiator systems. It then became apparent that it did not produce equally comfortable conditions, and that if it was to satisfy the occupants much greater care than had at first been thought necessary would have to be taken over the methods of introducing the warm air into occupied rooms.

But every refinement made has added a further restraint on the planning of the building. In the last couple of years we have got to the stage where, in order to make warm air heating work well, we have had to make so many demands of the architect that he is now beginning to ask us to go back to radiators.

I have no doubt at all that the system developed by the authors is a great improvement in warm air heating. Unfortunately it has the same side effect as every previous improvement had, namely it reduces the competitiveness of warm air in comparison with radiators.

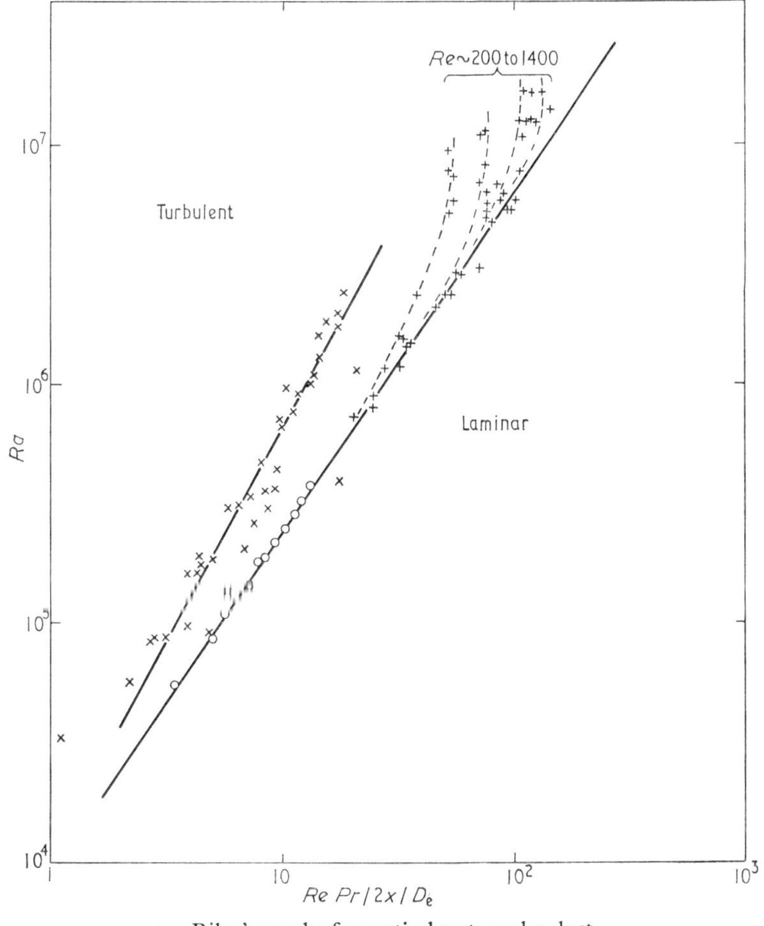

+ Riley's results for vertical rectangular duct.
× Hallman's results for vertical tubes.
○ Scheele *et al.* results for vertical tubes.

Fig. D2. A plot of the unmodified Rayleigh number at transition against the inverse of the dimensionless distance up the channel

J. A. Riley Member

With reference to Paper C112/71: The authors have extended transition theory to the case of developing upflow within an annulus. I wish to raise two points, both based upon a comparison of the present paper with my own previously published work on transition in upflow in a vertical rectangular duct [Reference (4)].

(i) The penultimate paragraph in the section headed 'Experimental Heat Transfer' tends to imply that there will be an insignificant gain in heat transfer during the transition to unstable flow and probably also gives the impression that this is the case with established unstable flow. This implication is contrary to the results obtained by me when investigating the transition phenomena in a vertical rectangular duct (4). Fig. 4 of reference (4) shows a distinct and very significant reduction in the channel wall temperature during transition and Fig. 9 of the same reference shows the Nusselt number magnitudes and variation in this region. For these tests, most of which were performed with relatively large Rayleigh numbers, the established unstable flow was, by visual observation within the channel and by channel wall temperature fluctuations, clearly turbulent, and was so stated by me in reference (4).

(ii) Hallman, reference (5), plots his transition data for upflow in vertical tubes together with the comparable data of reference (6) on his Fig. 14, using as one parameter the unmodified Rayleigh number. I have replotted these results in Fig. D2, together with my results obtained from the vertical rectangular duct experiments (4), using in my case as the characteristic dimension the duct equivalent diameter (taken as twice the width between the duct heated walls). Fluid properties are based on the local film temperature.

Reference to Fig. D2 indicates that using this equivalent diameter makes my results compatible with the results of references (5) and (6). The authors might therefore consider it worthwhile to adapt their transition data obtained with an annulus to obtain a similar comparison.

Notation. As used in reference (4).

REFERENCES

(4) Riley, J. A. 'Combined convective heat transfer in a vertical channel of rectangular cross section', *J. Br. nucl. Energy Soc.*, 1970 (Oct.) **9** (No. 4), 235.

(5) Hallman, T. M. 'Experimental study of combined forced and free laminar convection in a vertical tube', 1961, *National Aeronautics and Space Administration Report* TND-1104.

(6) Scheele, Rosen and Hanratty. 'Effect of natural convection on transition to turbulence in vertical pipes', *Can. J. chem. Engng.* 1960 (June) **38** (No. 3), 67.

Authors' Replies

K. Sherwin and J. D. Wallis

The datum chosen for temperature levels within our theoretical and experimental studies is the temperature difference between the outer and inner walls of the annulus. This was chosen as being measurable, and also provides a similar temperature datum to most other workers within the field. The validity of our choice of temperature datum is indicated by the good agreement between theoretical and experimental flow behaviour. Dr Allen's belief in the arithmetic mean temperature of the fluid as being the correct datum is probably equally valid, but it in no way proves that our datum is unsuitable, since the arithmetic mean temperature is also related to the wall temperature.

Concerning Mr Riley's first point, we stated that there was no marked change in the heat transfer performance during the transition to unstable flow behaviour. Our observations indicate that there was a steady improvement in the heat transfer coefficient throughout the steady region and extending into the unsteady region. This is in contrast to the case of buoyancy opposing flow, where flow changes occur near the heated wall, and there is a marked improvement in heat transfer performance.

To adapt our data to provide comparison with the work of others is only possible if the flow characteristics of the various systems are comparable. Unfortunately the annulus case cannot be compared with circular tubes, since velocity distributions are different. In the case of the annulus, the velocity is zero at the inner surface, while the velocity at the centre of a tube, or of any other open duct, is finite.

A. Lichtarowicz

The following correction to Paper C114 should be noted. Page 15, Fig. 114.5, values of *Nu*; for '3·0, 4·0, 5·0' read '4·0, 4·5, 5·0'.

M. W. Collins

The following correction to Paper C115 should be noted. Page 20, right-hand column, '**Effect of heat flux**', line 2; delete sentence beginning 'For comparative purposes . . .'.

B. H. Shaw

The experimental set-up used in this project, being an actual situation, was not ideal in layout. It was therefore thought best to use the generalized theory for natural convection, as put forward by Brown and Solvason in reference (3) on page 39, and to modify this in terms of combined natural convection and forced air flow.

Mr Porges' criticism of Brown and Solvason's theory, that when the enclosure is sealed there is no net flow of air across the opening under natural convection, and his query as to whether the absolute pressure is equal at the centre line of the opening, are best answered by the statements of Brown and Solvason. They consider that these assumptions are for the limiting case for density differences that are small compared with the mean density. The ensuing error in derivation of the heat and mass transfer equations will be negligible, except for gases at very low temperatures and for very large temperature differentials. I agree with these statements.

With regard to the question of the position of the line of equal pressure, or neutral zone, Fig. D3 shows the actual results; (*a*) represents a small temperature differential in the order of 0·11 degC, while (*b*) represents a larger differential of 5·59 degC. It may be seen from these results that the neutral zone does in fact exist approximately half way up the opening. It is also interesting to note that this occurs at the point of inflection of both the velocity profile and temperature gradient within the doorway. The inside and outside temperature gradients are also indicated. Although all my results are not as good as those presented, in fact, allowing for the screening of one door by air passing through the other, and a tendency at times for air to stream out at the corners, the point of inflection was found to be at the mid-point of the opening.

The best explanation for the rise in the coefficient at low temperature differentials is still the fact that ventilation in a room creates turbulence and air movement in the range 0·1016–0·1524 m/s, and thus causes an exchange of air through an opening, even with no temperature differential across it.

The second question deals with the assumption that there is no temperature gradient in either room. This assumption was not stated in the paper, and the author is indeed aware that gradients will exist. This is evident from Fig. D3, especially (*b*) which shows a gradient of almost 6 degC in one of the rooms. Even with such a large gradient there is no obvious effect on the shifting of the position of the neutral zone as Mr Porges thought may happen. The temperature differential used in the analysis of the results was that of the temperature difference between the top and bottom of the opening, and this was thought by the author to be the most appropriate differential with respect to the theory.

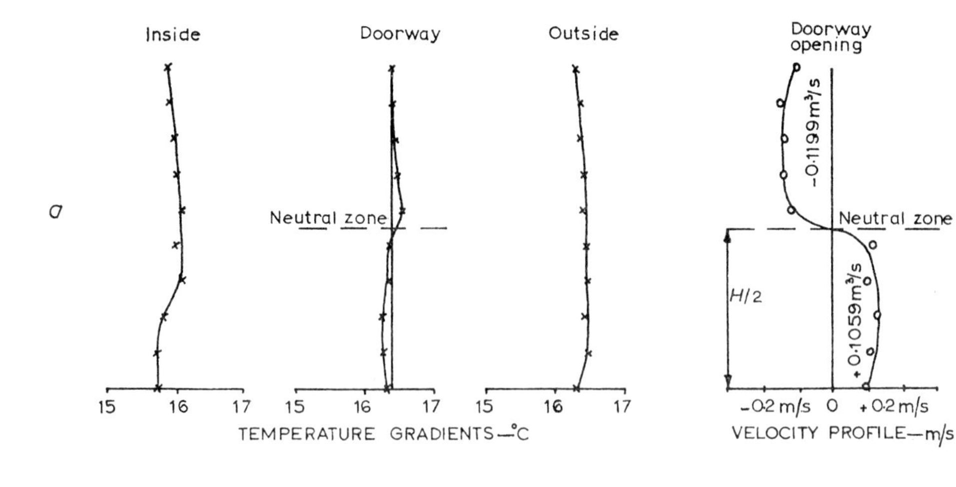

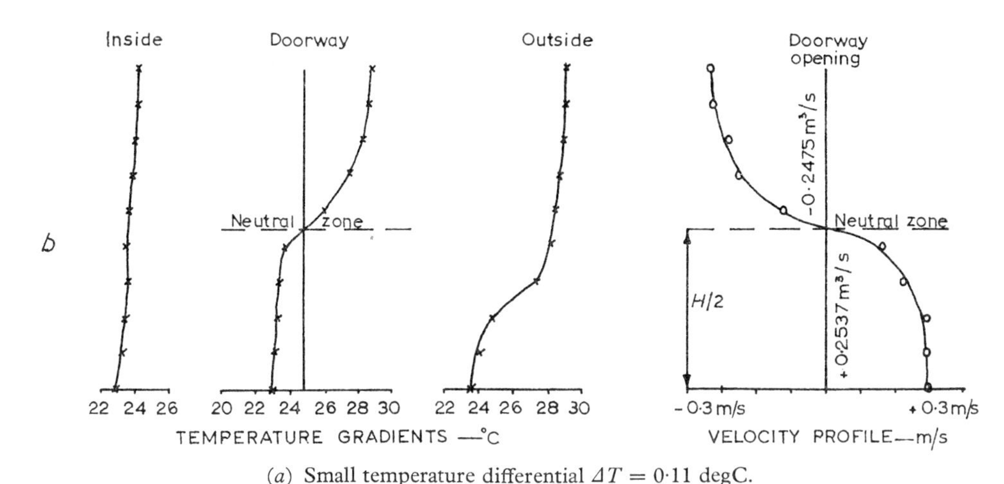

(*a*) Small temperature differential $\Delta T = 0{\cdot}11$ degC.
(*b*) Large temperature differential $\Delta T = 5{\cdot}59$ degC.

Fig. D3. Temperature gradients and velocity profiles indicating position

The following corrections to the paper should be noted.

Page 32, Notation reads Q_L, Q_x Leakage Transfer . . .; Q_x should be omitted.

Page 32, Dimensionless groups reads *Pr* Prandtl number [$= c_p U/k$]; U should be μ.

Page 33, equation (117.12) reads $P_2 = P_0 + \rho_2 zZ$; z should be g.

Page 36, Fig. 117.5, vertical ordinate reads $hH/C\rho\mu$; should be $c_p\mu$.

Page 38, Fig. 117.7, caption reads ($10^3 < Sw < 10^4$); should be ($10^3 < Sw < 10^6$).

Page 38, Reference (8) reads Houille blanche 1963; should be La Houille Blanche 1963. Fig. 117.7, vertical ordinate reads $C_\rho\mu$; should be $c_p\mu$.

J. E. Byrne and E. Ejiogu

The following correction to Paper C118 should be noted. Page 42, left-hand column, equation (118.9); for '*Rr*' read '*Gr*'.

List of Delegates

Name	Affiliation
Adebiyi, G. A.	University of Manchester.
Adelekan, W.	University of Manchester.
Allen, P. H. G.	Imperial College, London.
Andrews, T.	Charles Andrews and Sons, Sale, Cheshire.
Arnold, M. J.	The Gas Council, London.
Arscott, J. A.	Central Electricity Generating Board, London.
Bailey, D.	Electricity Council Research Centre, Chester.
Barrow, H.	University of Liverpool.
Blackbourn, M.	G.E.C. Reactor Equipment Ltd, Leicester.
Blaylock, G.	C. A. Parsons Ltd, Newcastle upon Tyne.
Brown, R.	U.W.I.S.T., Cathays Park, Cardiff.
Byrne, J. E. M.	University of Manchester.
Chisholm, D.	Applied Heat Division, National Engineering Laboratory, East Kilbride, Scotland.
Chullabodhi, C.	U.M.I.S.T., Manchester.
Clark, R. P.	National Institute for Medical Research, London.
Clements, A. A.	Central Electricity Generating Board, London.
Collins, M. W.	The City University, London.
Corlett, R.	Y-Ard Ltd, Glasgow.
Dampster, J.	Fluor (England) Ltd, London.
Eagle, G. R.	Mars Ltd, Slough, Bucks.
Ejiogu, E. U.	Foster Wheeler John Brown Boilers Ltd, London.
Elston, M. J. H.	Berkeley Nuclear Laboratories, Berkeley, Glos.
Etherington, C.	U.K.A.E.A. Reactor Group, Preston, Lancs.
Fewster, J.	University of Manchester.
Furber, B. N.	The Nuclear Power Group Ltd, Knutsford, Cheshire.
Gabrielides, A. G.	G. A. Gabrielides Ltd, Nicosia, Cyprus.
Green, C. H.	The Nuclear Power Group Ltd, Knutsford, Cheshire.
Green, P. K.	Post Office Telecommunications, Leeds.
Haddock, A. K.	Head Wrightson Ltd, Teesside.
Hall, W. B.	University of Manchester.
Harrington, E. L.	Dept. of Trade and Industry, Warrington, Lancs.
Hatton, A. P.	U.M.I.S.T., Manchester.
Hayward, A. R.	Sealed Motor Construction Ltd, Bridgwater, Somerset.
Higginbottom, D.	Berkeley Nuclear Laboratories, Berkeley, Glos.
Howarth, A. T.	Oxford Polytechnic, Headington, Oxford.
Jackson, J. D.	University of Manchester.
James, D. D.	U.M.I.S.T., Manchester.
Lacey, P. M. C.	University of Exeter.
Lichtarowicz, A.	University of Nottingham.
Loy, A. W.	U.M.I.S.T., Manchester.
Marshall, A.	C.E.G.B., Southampton.
Mynett, J. A.	University of Salford.
Ogunba, V. O.	University of Liverpool.
Padgham, F. K.	Pirelli General Ltd, Eastleigh, Hants.
Parker, E.	University of Salford.
Porges, F	John Porges (Consulting Engineers), London.
Price, P. H.	University of Manchester.
Rafique, S. O.	Engineering and Product Development, Covrad Ltd, Coventry.
Ralph, J. C.	A.E.R.E., Harwell, Berkshire.
Riley, J. A.	Scottish Research Reactor Centre, Glasgow.
Scott, P. A. J.	Current Product Development Dept, Covrad Ltd, Coventry.
Shaw, B. H.	Building Services Research Unit, Glasgow.
Sheppard, M.	G.E.C. Reactor Equipment Ltd, Leicester.
Sherratt, A. F. C.	University of Nottingham.
Sherwin, K.	University of Liverpool.
Simpson, H. S.	The Nuclear Power Group Ltd, Knutsford, Cheshire.
Smith, R. T.	Thomas Potterton Ltd, London.
Spence, I. D.	Reactor Equipment Ltd, Whetstone, Leicester.
Tate, C. J.	Liverpool Polytechnic.
Todd, J. P. H.	Central Electricity Generating Board, London.
Toy, N.	The City University, London.
Tudhope, R. G.	University of Strathclyde, Glasgow.
Watson, A.	University of Manchester.
Weinberg, R. S.	Simon Engineering Laboratories, Manchester.
White, W. J.	British Nuclear Design and Construction Whetstone, Leicester.
Wilkie, D.	U.K.A.E.A., Seascale, Cumberland.
Wolley, D. E.	Potterton International, London.
Woolley, N. H.	U.M.I.S.T., Manchester.

Index to Authors and Participants

Names of authors and numbers of pages on which papers begin are in bold type.

Subject Index

Titles of papers are in capital letters.

MADE AND PRINTED IN GREAT BRITAIN BY WILLIAM CLOWES & SONS, LIMITED, LONDON, BECCLES AND COLCHESTER

THE UNIVERSITY OF STRATHCLYDE
ANDERSONIAN
LIBRARY

THE UNIVERSITY OF STRATHCLYDE
ANDERSONIAN
LIBRARY